Revise AS

OCR Mathematics

Peter Sherran

Contents

	Specification information	4
	Your AS/A2 Level Mathematics course	5
	Exam technique	8
	What grade do you want?	10
	Four steps to successful revision	12

Chapter 1 Core 1 Pure Mathematics

1.1	Algebra and functions	13
1.2	Coordinate geometry	23
1.3	Differentiation	28
	Sample questions and model answers	32
	Practice examination questions	34

Chapter 2 Core 2 Pure Mathematics

2.1	Algebra and functions	36
2.2	Sequences and series	40
2.3	Trigonometry	44
2.4	Integration	49
	Sample questions and model answers	53
	Practice examination questions	55

Chapter 3 Mechanics 1

3.1	Kinematics	58
3.2	Statics	62
3.3	Dynamics	66
	Sample questions and model answers	69
	Practice examination questions	71

Chapter 4 Probability and Statistics 1

4.1	Representing data	75
4.2	Probability	78
4.3	Discrete random variables	81
4.4	Correlation and regression	84
	Sample questions and model answers	86
	Practice examination questions	88

Chapter 5 Decision Mathematics 1

5.1	Algorithms	91
5.2	Graphs and networks	94
5.3	Linear programming	99
	Sample questions and model answers	102
	Practice examination questions	104
	Practice examination answers	107
	Index	119

Specification information

OCR Mathematics

MODULE	SPECIFICATION TOPIC	CHAPTER REFERENCE	STUDIED IN CLASS	REVISED	PRACTICE QUESTIONS
Core 1	Indices and surds	1.1			
	Polynomials	1.1			
	Coordinate geometry and graphs	1.1, 1.2			
	Differentiation	1.3			
Core 2	Trigonometry	2.3			
	Sequences and series	2.2			
	Algebra	2.1			
	Integration	2.4			
Mechanics 1	Force as a vector	3.1, 3.2			
	Equilibrium of a particle	3.2			
	Kinematics of motion in a straight line	3.1			
	Newton's laws of motion	3.3			
	Linear momentum	3.3			
Probability and Statistics 1	Representation of data	4.1			
	Probability	4.2			
	Discrete random variables	4.3			
	Bivariate data	4.4			
Decision 1	Algorithms	5.1			
	Graph theory	5.2			
	Networks	5.2			
	Linear programming	5.3			

Examination analysis

The *assessment is by written papers. All questions are compulsory.*
Core 1 + Core 2 + one of Mechanics 1, Probability and Statistics 1 and Decision 1.

Core 1	AS	No calculator	1 hr 30 min exam	33.3%
Core 2	AS	Scientific/graphic calculator	1 hr 30 min exam	33.3%
Mechanics 1	AS	Scientific/graphic calculator	1 hr 30 min exam	33.3%
Probability and Statistics 1	AS	Scientific/graphic calculator	1 hr 30 min exam	33.3%
Decision 1	AS	Scientific/graphic calculator	1 hr 30 min exam	33.3%

Your AS/A2 Level Mathematics course

AS and A2

The OCR A Level course is in two parts, with three separate modules in each part. Students first study the AS (Advanced Subsidiary) course. Some will then go on to study the second part of the A Level course, called A2. Advanced Subsidiary is assessed at the standard expected halfway through an A Level course: i.e., between GCSE and Advanced GCE. This means that AS and A2 courses are designed so that difficulty steadily increases:

- AS Mathematics builds from GCSE Mathematics
- A2 Mathematics builds from AS Mathematics.

How will you be tested?

Assessment units

For AS Mathematics, you will be tested by three assessment units. For the full A Level in Mathematics, you will take a further three units. AS Mathematics forms 50% of the assessment weighting for the full A Level.

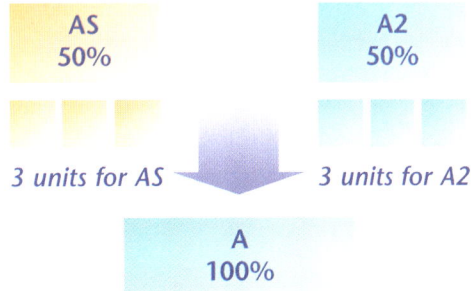

Each unit can normally be taken in either January or June. Alternatively, you can study the whole course before taking any of the unit tests. There is a lot of flexibility about when exams can be taken and the diagram below shows just some of the ways that the assessment units may be taken for AS and A Level Mathematics.

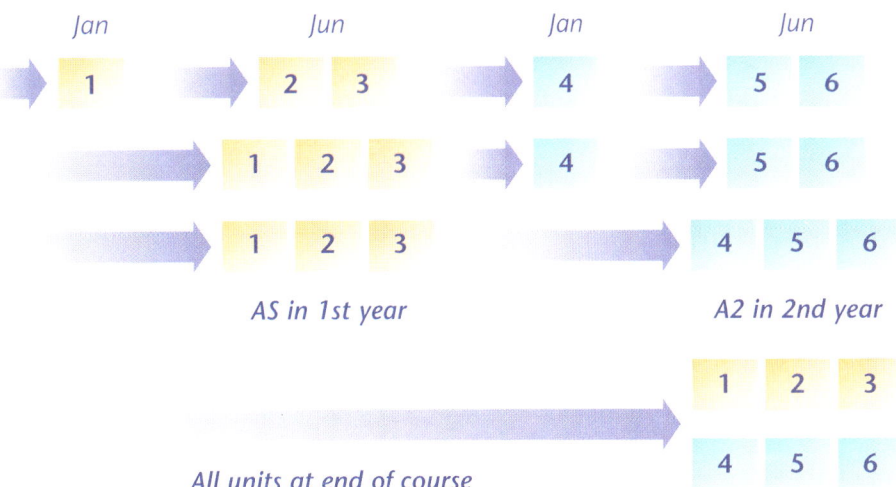

Your AS/A2 Level Mathematics course

If you are disappointed with a module result, you can resit the module. There is no restriction on the number of times a module may be attempted. The best available result for each module will count towards the final grade.

Synoptic assessment

The GCE Advanced Subsidiary and Advanced Level Qualification specific Criteria state that A Level specifications must include synoptic assessment (representing at least 20% of the total A Level marks).

Synoptic assessment in mathematics addresses candidates' understanding of the connections between different elements of the subject. It involves the explicit drawing together of knowledge, understanding and skills learned in different parts of the A Level course, through using and applying methods developed at earlier stages of the course to solving problems. Making and understanding connections in this way is intrinsic to learning mathematics.

Key skills

It is important that you develop your key skills throughout your AS and A2 courses. These are important skills that you need whatever you do beyond AS and A Levels. To gain the key skills qualification, which is equivalent to an AS Level, you will need to collect evidence together in a 'portfolio' to show that you have attained Level 3 in Communication, Application of number and Information technology. You will also need to take a formal test in each key skill. You will have many opportunities during AS Mathematics to develop your key skills.

It is a worthwhile qualification, as it demonstrates your ability to put your ideas across to other people, collect data and use up-to-date technology in your work.

What skills will I need?

For AS Mathematics, you will be tested by assessment objectives: these are the skills and abilities that you should have acquired by studying the course. The assessment objectives for AS Mathematics are shown below.

Candidates should be able to:

- recall, select and use their knowledge of mathematical facts, concepts and techniques in a variety of contexts
- construct rigorous mathematical arguments and proofs through use of precise statements, logical deduction and inference and by the manipulation of mathematical expressions, including the construction of extended arguments for handling substantial problems presented in unstructured form
- recall, select and use their knowledge of standard mathematical models to represent situations in the real world; recognise and understand given representations involving standard models; present and interpret results from such models in terms of the original situation, including discussion of the assumptions made and refinement of such models
- comprehend translations of common realistic contexts into Mathematics; use the results of calculations to make predictions, or comment on the context; and, where appropriate, read critically and comprehend longer mathematical arguments or examples of applications

Your AS/A2 Level Mathematics course

- use contemporary calculator technology and other permitted resources (such as formulae booklets or statistical tables*) accurately and efficiently; understand when not to use such technology, and its limitations; give answers to appropriate accuracy.

* You can find a copy of the formulae booklet and statistical tables in the A Level mathematics section of the OCR website, www.ocr.org.uk

Progression and prior learning

Mathematics is, inherently, a sequential subject. There is a progression of material through all levels at which the subject is studied. The criteria therefore build on the knowledge, understanding and skills established at GCSE.

Thus, candidates embarking on AS/A2 Level study in Mathematics subjects are expected to have covered all the material in the Higher Tier. This material is regarded as assumed background knowledge. However, it may be assessed within questions focused on other material from the specification.

Exam technique

What are the examiners looking for?

Examiners use certain words in their instructions to let you know what they are expecting in your answer. Make sure that you know what they mean so that you can give the right response.

Write down, state

You can write your answer without having to show how it was obtained. There is nothing to prevent you doing some working if it helps you, but if you are doing a lot then you might have missed the point.

Calculate, find, determine, show, solve

Make sure that you show enough working to justify the final answer or conclusion. Marks will be available for showing a correct method.

Deduce, hence

This means that you are expected to use the given result to establish something new. You must show all of the steps in your working.

Draw

This is used to tell you to plot an accurate graph using graph paper. Take note of any instructions about the scale that must be used. You may need to read values from your graph.

Sketch

If the instruction is to sketch a graph then you don't need to plot the points but you will be expected to show its general shape and its relationship with the axes. Indicate the positions of any turning points and take particular care with any asymptotes.

Find the exact value

This instruction is usually given when the final answer involves an irrational value such as a logarithm, π or a surd. You will need to demonstrate that you can manipulate these quantities so don't just key everything into your calculator or you will lose marks.

If a question requires the final answer to be given to a specific level of accuracy then make sure that you do this or you might needlessly lose marks.

Exam technique

Some dos and don'ts

Dos

Do read the question

- Make sure that you are clear about what you are expected to do. Look for some structure in the question that may help you take the right approach.
- Read the question *again* after you have answered it as a quick check that your answer is in the expected form.

Do use diagrams

- In some questions, particularly in mechanics, a clearly labelled diagram is essential. Use a diagram whenever it may help you understand or represent the problem that you are trying to solve.

Do take care with notation

- Write clearly and use the notation accurately. Use brackets when they are required.
- Even if your final answer is wrong, you may earn some marks for a correct expression in your working.

Do avoid silly answers

- Check that your final answer is sensible within the context of the question.

Do make good use of time

- Choose the order in which you answer the questions carefully. Do the ones that are easiest for you first.
- Set yourself a time limit for a question depending on the number of marks available.
- Be prepared to leave a difficult part of a question and return to it later if there is time.
- Towards the end of the exam make sure that you pick up all of the easy marks in any questions that you haven't got time to answer fully.

Don'ts

Don't work with rounded values

- There may be several stages in a solution that produce numerical values. Rounding errors from earlier stages may distort your final answer. One way to avoid this is to make use of your calculator memories to store values that you will need again.

Don't cross out work that may be partly correct

- It's tempting to cross out something that hasn't worked out as it should. Avoid this unless you have time to replace it with something better.

Don't write out the question

- This wastes time. The marks are for your solution!

What grade do you want?

Everyone should be able to improve their grades but you will only manage this with a lot of hard work and determination. The details given below describe a level of performance typical of candidates achieving grades A, C or E. You should find it useful to read and compare the expectations for the different levels and to give some thought to the areas where you need to improve most.

Grade A candidates

- Recall or recognise almost all the mathematical facts, concepts and techniques that are needed, and select appropriate ones to use in a variety of contexts.
- Manipulate mathematical expressions and use graphs, sketches and diagrams, all with high accuracy and skill.
- Use mathematical language correctly and proceed logically and rigorously through extended arguments or proofs.
- When confronted with unstructured problems they can often devise and implement an effective solution strategy.
- If errors are made in their calculations or logic, these are sometimes noticed and corrected.
- Recall or recognise almost all the standard models that are needed, and select appropriate ones to represent a wide variety of situations in the real world.
- Correctly refer results from calculations using the model to the original situation; they give sensible interpretations of their results in the context of the original realistic situation.
- Make intelligent comments on the modelling assumptions and possible refinements to the model.
- Comprehend or understand the meaning of almost all translations into mathematics of common realistic contexts.
- Correctly refer the results of calculations back to given context and usually make sensible comments or predictions.
- Can distil the essential mathematical information from extended pieces of prose having mathematical content.
- Comment meaningfully on the mathematical information.
- Make appropriate and efficient use of contemporary calculator technology and other permitted resources, and are aware of any limitations to their use.
- Present results to an appropriate degree of accuracy.

Grade C candidates

- Recall or recognise most of the mathematical facts, concepts and techniques that are needed, and usually select appropriate ones to use in a variety of contexts.
- Manipulate mathematical expressions and use graphs, sketches and diagrams, all with a reasonable level of accuracy and skill.
- Use mathematical language with some skill and sometimes proceed logically through extended arguments or proofs.
- When confronted with unstructured problems they sometimes devise and implement an effective and efficient solution strategy.
- Occasionally notice and correct errors in their calculations.
- Recall or recognise most of the standard models that are needed and usually select appropriate ones to represent a variety of situations in the real world.

What grade do you want?

- Often correctly refer results from calculations using the model to the original situation; they sometimes give sensible interpretations of their results in context of the original realistic situation.
- Sometimes make intelligent comments on the modelling assumptions and possible refinements to the model.
- Comprehend or understand the meaning of most translations into mathematics of common realistic contexts.
- Often correctly refer the results of calculations back to the given context and sometimes make sensible comments or predictions.
- Distil much of the essential mathematical information from extended pieces of prose having mathematical content.
- Give some useful comments on this mathematical information.
- Usually make appropriate and effective use of contemporary calculator technology and other permitted resources, and are sometimes aware of any limitations to their use.
- Usually present results to an appropriate degree of accuracy.

Grade E candidates

- Recall or recognise some of the mathematical facts, concepts and techniques that are needed, and sometimes select appropriate ones to use in some contexts.
- Manipulate mathematical expressions and use graphs, sketches and diagrams, all with some accuracy and skill.
- Sometimes use mathematical language correctly and occasionally proceed logically through extended arguments or proofs.
- Recall or recognise some of the standard models that are needed and sometimes select appropriate ones to represent a variety of situations in the real world.
- Sometimes correctly refer results from calculations using the model to the original situation; they try to interpret their results in the context of the original realistic situation.
- Sometimes comprehend or understand the meaning of translations into mathematics of common realistic contexts.
- Sometimes correctly refer the results of calculations back to the given context and attempt to give comments or predictions.
- Distil some of the essential mathematical information from extended pieces of prose having mathematical content; they attempt to comment on this mathematical information.
- Candidates often make appropriate and efficient use of contemporary calculator technology and other permitted resources.
- Often present results to an appropriate degree of accuracy.

The table below shows how your uniform standardised mark is translated.

average	80%	70%	60%	50%	40%
grade	A	B	C	D	E

To achieve an A* grade in Mathematics, you need to achieve a grade A overall (an average of 80 or more on uniform mark scale) for the whole A Level qualification and an average of 90 or more on the uniform mark scale in Core 3 and Core 4. It is awarded for A Level qualification only and not for the AS qualification or individual units.

Four steps to successful revision

Step 1: Understand

- Study the topic to be learned slowly. Make sure you understand the logic or important concepts.
- Mark up the text if necessary – underline, highlight and make notes.
- Re-read each paragraph slowly.

GO TO STEP 2

Step 2: Summarise

- Now make your own revision note summary:
  What is the main idea, theme or concept to be learned?
  What are the main points? How does the logic develop?
  Ask questions: Why? How? What next?
- Use bullet points, mind maps, patterned notes.
- Link ideas with mnemonics, mind maps, crazy stories.
- Note the title and date of the revision notes
  (e.g. Mathematics: Trigonometry, 3rd March).
- Organise your notes carefully and keep them in a file.

This is now in **short-term memory**. You will forget 80% of it if you do not go to Step 3.
GO TO STEP 3, but first take a 10 minute break.

Step 3: Memorise

- Take 25 minute learning 'bites' with 5 minute breaks.
- After each 5 minute break test yourself:
  Cover the original revision note summary
  Write down the main points
  Speak out loud (record yourself)
  Tell someone else
  Repeat many times.

The material is well on its way to **long-term memory**.
You will forget 40% if you do not do step 4. GO TO STEP 4

Step 4: Track/Review

- Create a Revision Diary (one A4 page per day).
- Make a revision plan for the topic, e.g. 1 day later, 1 week later, 1 month later.
- Record your revision in your Revision Diary, e.g.
  Mathematics: Trigonometry, 3rd March 25 minutes
  Mathematics: Trigonometry, 5th March 15 minutes
  Mathematics: Trigonometry, 3rd April 15 minutes
  ... and then at monthly intervals.

Chapter 1
Core 1 Pure Mathematics

The following topics are covered in this chapter:

- Algebra and functions
- Coordinate geometry
- Differentiation

Note: Calculators are *not* allowed in this module

1.1 Algebra and functions

After studying this section you should be able to:

- work with indices and surds
- use function notation
- solve quadratic equations and sketch the graphs of quadratic functions
- understand the definition of a polynomial
- recognise and sketch graphs of a range of functions
- use transformations to sketch graphs of related functions
- solve simultaneous equations – one linear and one quadratic
- solve linear and quadratic inequalities

Indices

You need to know these basic rules and be able to apply them.

$$a^m \times a^n = a^{m+n} \qquad a^m \div a^n = a^{m-n} \qquad (a^m)^n = a^{mn} \qquad a^{\frac{1}{n}} = \sqrt[n]{a}$$

$$a^{\frac{m}{n}} = (a^m)^{\frac{1}{n}} = (a^{\frac{1}{n}})^m \qquad a^{-n} = \frac{1}{a^n} \qquad (ab)^n = a^n b^n \qquad \left(\frac{a}{b}\right)^n = \frac{a^n}{b^n}$$

Remember that, in general $(a+b)^n \neq a^n + b^n$ and $(a-b)^n \neq a^n - b^n$.

Some important special cases are: $a^1 = a \qquad a^{\frac{1}{2}} = \sqrt{a}$

and $a^0 = 1 \qquad a^{-1} = \dfrac{1}{a}$ provided $a \neq 0$.

For example $\quad 9^{\frac{1}{2}} = \sqrt{9} = 3 \quad$ and $\quad 16^{\frac{3}{2}} = (16^{\frac{1}{2}})^3 = 4^3 = 64$.

Surds

A **surd** is the root of a whole number that has an **irrational** value.

Some examples are $\sqrt{2}, \sqrt{3}$ and $\sqrt{10}$.

You can often simplify a surd using the fact that $\sqrt{ab} = \sqrt{a} \times \sqrt{b}$ and choosing a or b to be a square number.

An irrational number continues for ever after the decimal point without a repeating pattern.

For example $\sqrt{12} = \sqrt{4 \times 3} = \sqrt{4} \times \sqrt{3} = 2\sqrt{3}$ and $\sqrt{18} + \sqrt{32} = 3\sqrt{2} + 2 \times 4\sqrt{2} = 11\sqrt{2}$.

To simplify $\dfrac{3}{\sqrt{5}}$, multiply the numerator and the denominator by $\sqrt{5}$ to get

$\dfrac{3}{\sqrt{5}} \times \dfrac{\sqrt{5}}{\sqrt{5}} = \dfrac{3\sqrt{5}}{5}$. This is called **rationalising the denominator**.

13

Core 1 Pure Mathematics

Functions

A **function** may be thought of as a rule which takes each member x of a set and assigns, or **maps**, it to some value y known as its **image**.

x maps to y
y is the image of x.

$$x \longrightarrow \boxed{\text{Function}} \longrightarrow y$$

A letter such as f, g or h is often used to stand for a function. The function which squares a number and adds on 5, for example, can be written as $f(x) = x^2 + 5$. The same notation may also be used to show how a function affects particular values. For this function, $f(4) = 4^2 + 5 = 21$, $f(-10) = (-10)^2 + 5 = 105$ and so on.

$f(x)$ is read as 'f of x'.

An alternative notation for the same function is $f: x \mapsto x^2 + 5$.

Quadratics

In algebra, any expression of the form $ax^2 + bx + c$ where $a \neq 0$ is called a **quadratic expression**.

You need to be able to expand brackets, **for example**

$$(2x + 3)(x - 4) = 2x^2 - 8x + 3x - 12$$
$$= 2x^2 - 5x - 12$$

It is useful to remember these special results:

$$(x + a)^2 = x^2 + 2ax + a^2 \qquad (x - a)^2 = x^2 - 2ax + a^2 \qquad (x + a)(x - a) = x^2 - a^2$$

Some examples are:

$$(x + 3)^2 = x^2 + 6x + 9 \qquad\qquad (x - 5)^2 = x^2 - 10x + 25$$
$$(x + 7)(x - 7) = x^2 - 49 \qquad\qquad (x + \sqrt{3})(x - \sqrt{3}) = x^2 - 3$$

To solve problems in algebra you need to develop your skills so that you can recognise how to apply the basic results.

In this fraction, the denominator $2 - \sqrt{3}$ is irrational.

This is an example of rationalising the denominator.

For example, the fraction $\dfrac{1}{2 - \sqrt{3}}$ may be simplified by multiplying the numerator and denominator by $2 + \sqrt{3}$ to make an equivalent fraction in which the denominator is rational.

$(2 - \sqrt{3})(2 + \sqrt{3})$
$= 2^2 - (\sqrt{3})^2$
$= 4 - 3$.

$$\dfrac{1}{2 - \sqrt{3}} = \dfrac{1}{2 - \sqrt{3}} \times \dfrac{2 + \sqrt{3}}{2 + \sqrt{3}}$$

$$= \dfrac{2 + \sqrt{3}}{4 - 3} \qquad \text{(The denominator is now rational)}$$

$$= 2 + \sqrt{3}$$

Quadratic equations

Equations of the form $ax^2 + bx + c = 0$ (where $a \neq 0$) are **quadratic equations**.

Some quadratic equations can be solved by **factorising** the quadratic expression.

Core 1 Pure Mathematics

Example Solve $2x^2 - x - 3 = 0$.
$(2x - 3)(x + 1) = 0$
Either $2x - 3 = 0$ or $x + 1 = 0$
so $x = \frac{3}{2}$ or $x = -1$.

Factorise.

Some equations can be converted into a quadratic equation by substitution.

Example Solve $x^{\frac{2}{3}} + x^{\frac{1}{3}} - 6 = 0$
Substituting $y = x^{\frac{1}{3}}$ gives
$y^2 + y - 6 = 0$
$(y + 3)(y - 2) = 0$
Either $y + 3 = 0$ or $y - 2 = 0$
so $y = -3$ or $y = 2$.

Re-writing in terms of x gives $x^{\frac{1}{3}} = -3$ or $x^{\frac{1}{3}} = 2$,

so $x = -27$ or $x = 8$.

If the quadratic will not factorise then you can try **completing the square**.

$x^2 - 6x = (x - 3)^2 - 9$ so
$x^2 - 6x + 1 = (x - 3)^2 - 8$.

Example Solve $x^2 - 6x + 1 = 0$
$(x - 3)^2 - 8 = 0$ (Now x only appears once in the equation)
$(x - 3)^2 = 8$
$x - 3 = \pm \sqrt{8}$
$x = 3 \pm 2\sqrt{2}$.

$\sqrt{8} = \sqrt{(4 \times 2)} = 2\sqrt{2}$.

(The solutions are $3 + 2\sqrt{2}$ and $3 - 2\sqrt{2}$)

The method shown for completing the square can be adapted for the general form of a quadratic. This gives the **quadratic formula**.

$ax^2 + bx + c = 0$

Multiply both sides by $4a$.

$4a^2x^2 + 4abx + 4ac = 0$ Note: $4a^2x^2 + 4abx + b^2 = (2ax + b)^2$.

Add b^2 to complete the square and subtract it again to keep things the same.

$4a^2x^2 + 4abx + b^2 + 4ac - b^2 = 0$
$(2ax + b)^2 + 4ac - b^2 = 0$ Now x only appears once in the equation.
$(2ax + b)^2 = b^2 - 4ac$

Find the square root of both sides.

$2ax + b = \pm \sqrt{b^2 - 4ac}$

Rearrange to find x.

$x = \dfrac{-b \pm \sqrt{b^2 - 4ac}}{2a}$ You must learn this formula.

Example Solve $3x^2 - 4x - 2 = 0$ giving your answers in surd form.

Comparing this equation with the general form gives $a = 3$, $b = -4$ and $c = -2$. Substitute this information into the formula:

In Core 1 you do not have access to a calculator so you will have to leave the answer in surd form.

$x = \dfrac{4 \pm \sqrt{(-4)^2 - 4(3)(-2)}}{6} = \dfrac{4 \pm \sqrt{40}}{6} = \dfrac{4 \pm 2\sqrt{10}}{6} = \dfrac{2 \pm \sqrt{10}}{3}$

so $x = \dfrac{2 + \sqrt{10}}{3}$ or $x = \dfrac{2 - \sqrt{10}}{3}$.

Core 1 Pure Mathematics

Discriminant

In the formula, the value of $b^2 - 4ac$ is called the **discriminant**. This value can be used to give information about the solutions without having to solve the equation.

$b^2 - 4ac > 0$ **two** distinct real solutions (the solutions are often called **roots**)

In the example above, $b^2 - 4ac = 40 > 0$ and two real roots were found.

$b^2 - 4ac = 0$ **one** real solution (often thought of as a **repeated root**)

$b^2 - 4ac < 0$ **no** real solutions.

The solutions of a quadratic equation correspond to where the graph of the quadratic function crosses the x-axis. There are three possible situations depending on the value of the discriminant.

The diagrams correspond to the situation where $a > 0$. The same principle applies when $a < 0$ but the diagrams appear the other way up.

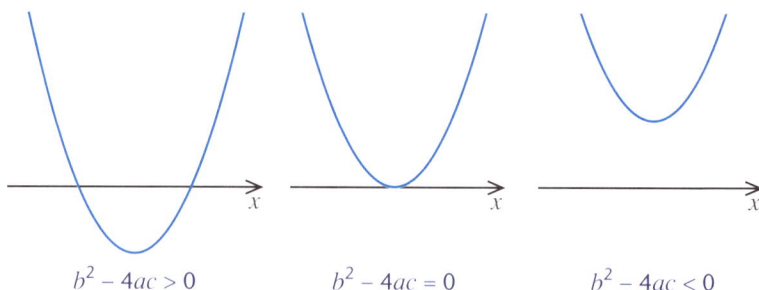

$b^2 - 4ac > 0$ $b^2 - 4ac = 0$ $b^2 - 4ac < 0$

Example   The equation $5x^2 + 3x + p = 0$ has a repeated root. Find the value of p.

In this case, $a = 5$, $b = 3$ and $c = p$.

For a repeated root $b^2 - 4ac = 0$ so $9 - 20p = 0$, giving $p = 0.45$

Quadratic graphs

The graph of $y = ax^2 + bx + c$ is a parabola.

$a > 0$ $a < 0$

The methods used for solving quadratic equations can also be used to give information about the graphs.

Example   Sketch the graph of $y = x^2 - x - 6$.
Find the coordinates of the lowest point on the curve.

The curve will cross the x-axis when $y = 0$. You can find these points by solving the equation $x^2 - x - 6 = 0$.

The curve will cross the y-axis when $x = 0$, giving $y = -6$.

$x^2 - x - 6 = 0 \Rightarrow (x + 2)(x - 3) = 0$

$\Rightarrow x = -2$ or $x = 3$

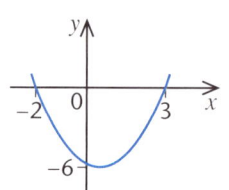

The curve is symmetrical so the lowest point occurs mid-way between −2 and 3 and this is given by (−2 + 3) ÷ 2 = 0.5

When $x = 0.5$, $y = 0.5^2 − 0.5 − 6 = −6.25$

The lowest point on the curve is (0.5, −6.25).

The vertex of the curve occurs at its maximum or minimum point.

Completing the square gives information about the **vertex** of the curve even if the equation will not factorise.

Example   Find the coordinates of the vertex of the curve $y = x^2 + 2x + 3$.

You need to recognise that $x^2 + 2x + 1 = (x + 1)^2$,

then, completing the square, $x^2 + 2x + 3 = x^2 + 2x + 1 + 2 = (x + 1)^2 + 2$.

The equation of the curve can now be written as $y = (x + 1)^2 + 2$.

$(x + 1)^2$ cannot be negative so its minimum value is zero, when $x = −1$.

This means that the minimum value of y is 2 and this occurs when $x = −1$.

The vertex of the curve is at (−1, 2).

Polynomials

A quadratic is one example of a **polynomial**. In general, a polynomial takes the form:

$$a_n x^n + a_{n-1} x^{n-1} + a_{n-2} x^{n-2} + \ldots + a_0,$$

This is much simpler than it looks.

where $a_n, a_{n-1}, \ldots a_0$ are constants and n is a positive whole number.

For example, $x^4 − 2x^3$ is a polynomial. In this case $a_4 = 1$, $a_3 = −2$ and a_2, a_1, a_0 are all zero.

The **degree** of a polynomial is the highest power of x that it includes, so the degree of $x^4 − 2x^3$ is 4. A quadratic is a polynomial of degree 2, a cubic is a polynomial of degree 3 and so on.

More graphs

The graph of a cubic function $y = ax^3 + bx^2 + cx + d$ can take a number of forms.

$a > 0$ $a < 0$

> **KEY POINT**
> The graph of a cubic function that can be factorised as $y = (x − p)(x − q)(x − r)$ will cross the x-axis at p, q and r. If any two of p, q and r are the same then the x-axis will be a tangent to the curve at that point.

Core 1 Pure Mathematics

For example, the graph of $y = (x+2)(x-3)^2$ looks like this:

This shows a repeated root of $(x+2)(x-3)^2 = 0$ at $x = 3$.

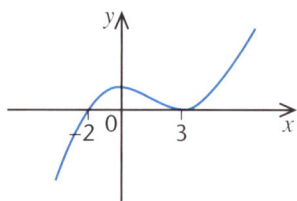

KEY POINT

The graph of $y = x^n$ where n is an integer has:
rotational symmetry about the **origin** when n is **odd**
reflective symmetry in the **y-axis** when n is **even**.

The graphs of $y = x^3$, $y = x^5$, $y = x^7$ … look something like this:

The graphs of $y = x^2$, $y = x^4$, $y = x^6$ … look something like this:

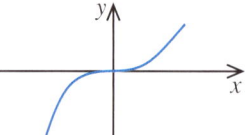
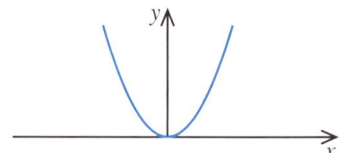

For higher powers of x the graphs are flatter between -1 and 1 and steeper elsewhere.

KEY POINT

A straight line that a graph approaches ever more closely without actually touching it is called an **asymptote**.

The graphs of $y = x^{-1}$, $y = x^{-3}$, $y = x^{-5}$ … look something like this:

The graphs of $y = x^{-2}$, $y = x^{-4}$, $y = x^{-6}$ … look something like this:

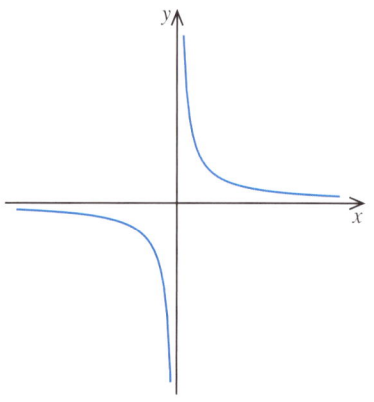

As the powers of x become more negative, the graphs become steeper between -1 and 1 and flatter elsewhere.

Both the x and y-axes are asymptotes for these graphs.

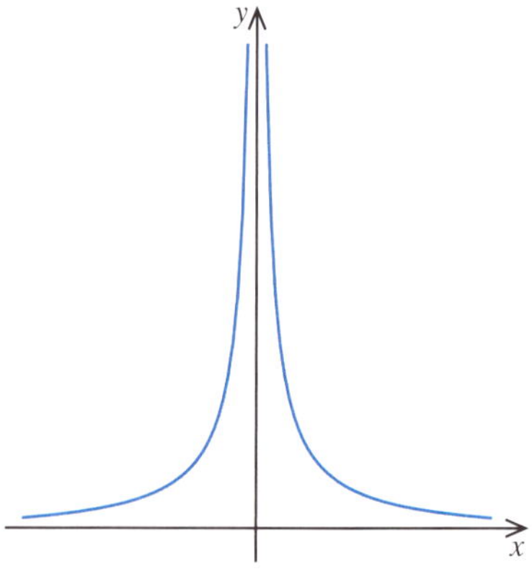

The x-axis and the positive y-axis are asymptotes for these graphs.

The graphs of $y = k\sqrt{x}$, $k > 0$, look something like this. These graphs do not have any asymptotes.

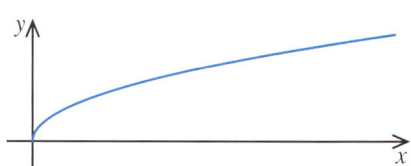

Core 1 Pure Mathematics

Transforming graphs

The graph of some new function can often be obtained from the graph of a known function by applying a transformation. A summary of the standard transformations is given in the table.

Known function	New function	Transformation
$y = f(x)$	$y = f(x) + a$	Translation through a units parallel to y-axis.
	$y = f(x - a)$	Translation through a units parallel to x-axis.
	$y = af(x)$	One-way stretch with scale factor a parallel to the y-axis.
	$y = f(ax)$	One-way stretch with scale factor $\frac{1}{a}$ parallel to the x-axis.

You may need to apply a combination of transformations in some cases.

Example The diagram shows the graph of a function, $y = f(x)$ for $1 \leqslant x \leqslant 3$

Use the same axes to show:
(a) $y = f(x) + 1$
(b) $y = f(x + 1)$
(c) $y = 2f(x)$
(d) $y = f(2x)$

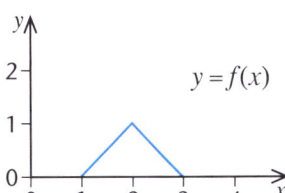

(a)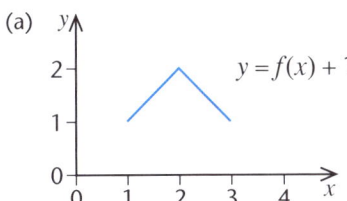

(b)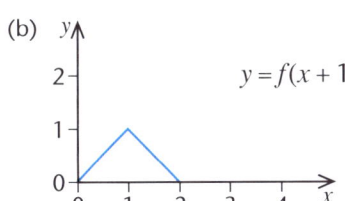

(c)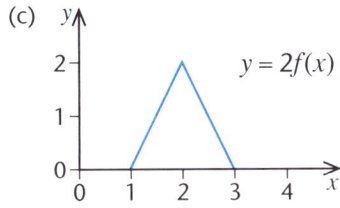

(d)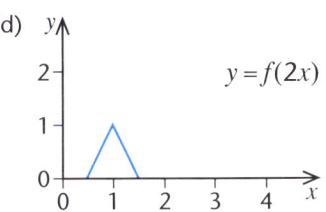

Simultaneous equations

The substitution method of solving pairs of linear simultaneous equations can also be applied when one of the equations is a quadratic.

Example Find the coordinates of the points where the line $y = x + 1$ intersects the circle $x^2 + y^2 = 5$.

$$y = x + 1 \quad (1)$$
$$x^2 + y^2 = 5 \quad (2)$$

Substitute for y from (1) into (2): $x^2 + (x + 1)^2 = 5.$

Now expand the brackets: $x^2 + x^2 + 2x + 1 = 5.$

Arrange in the form $ax^2 + bx + c = 0$: $2x^2 + 2x - 4 = 0.$

Divide by 2. $x^2 + x - 2 = 0$

19

Core 1 Pure Mathematics

Solve the equation: $(x - 1)(x + 2) = 0$
Either $x = 1$ or $x = -2$

Substitute into (1) to find the y values:
When $x = 1$, $y = 2$
When $x = -2$, $y = -1$

The coordinates of the points of intersection are (1, 2) and (−2, −1).

Geometrical interpretation of algebraic solutions

In the above example, when $y = x + 1$ and $x^2 + y^2 = 5$ are solved simultaneously, the resulting quadratic equation in x has two distinct solutions. This gives the two points of intersection of the line and the curve.

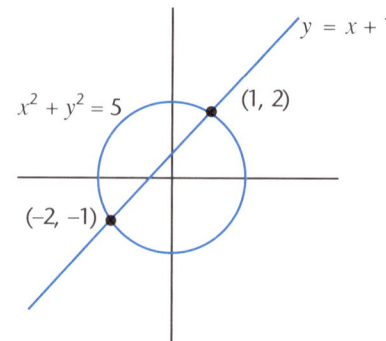

When you solve the equations of a line and a curve simultaneously and form a quadratic equation, $ax^2 + bx + c = 0$, the discriminant, $b^2 - 4ac$, gives information about the number of points of intersection.

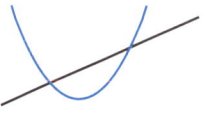 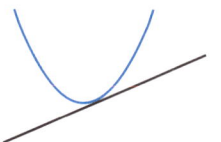 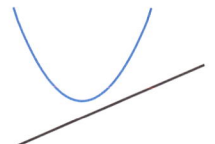

If $b^2 - 4ac > 0$ the line and curve intersect in two distinct points.

If $b^2 - 4ac = 0$ the line is a tangent to the curve.

If $b^2 - 4ac < 0$ the line and the curve do not intersect.

Inequalities

Linear inequalities can be solved by rearrangement in much the same way as linear equations. However, care must be taken to reverse the direction of the inequality when multiplying or dividing by a negative.

Example   Solve the inequality $8 - 3x > 23$.

Subtract 8 from both sides: $-3x > 15$
Divide both sides by −3: $x < -5$.

An inequality which has x on both sides is treated like the corresponding equation.

Example   Solve the inequality $5x - 3 > 3x - 10$.

Subtract $3x$ from both sides: $2x - 3 > -10$
Add 3 to both sides: $2x > -7$
Divide both sides by 2: $x > -3.5$

Quadratic inequalities are solved in a similar way to quadratic equations but a sketch graph is often helpful at the final stage.

Core 1 Pure Mathematics

Example Solve the inequality $x^2 - 3x + 2 < 0$.

Factorise the quadratic expression: $(x - 1)(x - 2) < 0$.

Sketch the graph of $y = (x - 1)(x - 2)$:

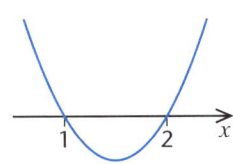

y = 0 when x = 1 or when x = 2. The graph must cross the x-axis at these points.

The graph shows that $(x - 1)(x - 2) < 0$ for x values between 1 and 2.

(x − 1)(x − 2) < 0 when the curve is below the x-axis.

It follows that $x^2 - 3x + 2 < 0$ when $1 < x < 2$.

If the quadratic expression cannot be factorised then the formula may be used to find the points of intersection of the curve with the x-axis.

Example Solve the inequality $x^2 + 2x - 5 > 0$.

Use the formula to solve $x^2 + 2x - 5 = 0$. $\quad x = \dfrac{-2 \pm \sqrt{24}}{2} = \dfrac{-2 \pm 2\sqrt{6}}{2}$

$\sqrt{24} = \sqrt{4 \times 6} = 2\sqrt{6}$

Simplify the result. $\quad x = -1 \pm \sqrt{6}$

Sketch the graph of $y = x^2 + 2x - 5$:

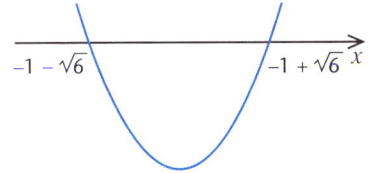

$x^2 + 2x - 5 > 0$ when the curve is above the x-axis.

Write the solution as two separate inequalities.

From the sketch, $x^2 + 2x - 5 > 0$ when $x < -1 - \sqrt{6}$ or when $x > -1 + \sqrt{6}$.

21

Core 1 Pure Mathematics

Progress check

1. Simplify these surd expressions:
   (a) $\sqrt{72}$ (b) $(1 + \sqrt{3})(1 - \sqrt{3})$ (c) $5\sqrt{12} - 6\sqrt{3}$.

2. (a) Solve $2x^2 + x - 21 = 0$ by factorising.
   (b) Solve $x^2 - 6x + 7 = 0$ by completing the square. Give the roots in surd form.
   (c) Solve $3x^2 + 2x - 2 = 0$ using the quadratic formula. Give the roots in surd form.

3. Express $x^2 - 8x + 17$ in the form $(x - a)^2 + b$ and hence write down its minimum value.

4. Sketch the graph of $y = (x + 3)(x - 2)^2$.

5. The diagram shows the graph of $y = f(x)$ for $0 \leq x \leq 2$. Sketch the graph of $y = f(x - 2)$.

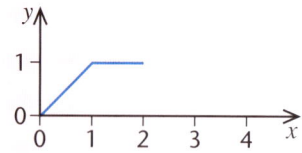

6. Solve these equations simultaneously.
   $x + 2y = 6$
   $4y^2 - 5x^2 = 36$.

7. Find the values of A, B and C.
   $3x^3 - 16x - 1 \equiv Ax(x - 1)(x + 2) + B(x - 1)^2 + C(4x^2 + 3)$.

8. Solve these inequalities.
   (a) $3x - 7 > 5x + 11$ (b) $2x^2 - 3x - 5 \leq 0$.

9. Solve $x^{\frac{2}{3}} - x^{\frac{1}{3}} - 12 = 0$.

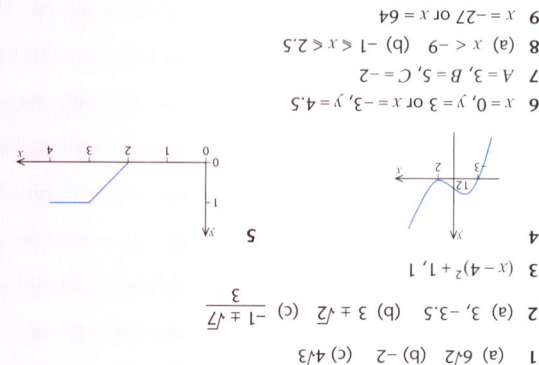

1. (a) $6\sqrt{2}$ (b) -2 (c) $4\sqrt{3}$
2. (a) $3, -3.5$ (b) $3 \pm \sqrt{2}$ (c) $\dfrac{-1 \pm \sqrt{7}}{3}$
3. $(x-4)^2 + 1$, 1
6. $x = 0, y = 3$ or $x = -3, y = 4.5$
7. $A = 3, B = 5, C = -2$
8. (a) $x > -9$ (b) $-1 \leq x \leq 2.5$
9. $x = -27$ or $x = 64$

1.2 Coordinate geometry

After studying this section you should be able to:

- find the gradient of a line joining two points
- recognise equations of straight lines in various forms
- construct the equation of a straight line parallel to a given line and passing through a given point
- construct the equation of a straight line perpendicular to a given line and passing through a given point
- construct the equation of a straight line passing through two given points
- find the coordinates of the mid-point of two given points
- find the distance between any two given points
- use the equation of a circle and circle properties

Gradient of a line

> **KEY POINT**
> The gradient of a line joining the points (x_1, y_1) and (x_2, y_2) is given by the formula $m = \dfrac{y_2 - y_1}{x_2 - x_1}$.

For example, the gradient of the line joining $(2, -5)$ and $(-1, 4)$ is
$$\dfrac{4 - (-5)}{-1 - 2} = \dfrac{9}{-3} = -3.$$

Straight lines

The general equation of a straight line is $ax + by + c = 0$. The equation of any straight line can be written in this form.

For example, the line $x + y = 5$ corresponds to $a = 1$, $b = 1$ and $c = -5$.

The gradient of a vertical line is undefined.

Straight lines, apart from those parallel to the y-axis, can be written in the form $y = mx + c$. This is known as **gradient–intercept** form because the gradient (m) and the y-intercept (c) are clearly identified in the equation. This makes it easy to construct an equation when the gradient and intercept are known.

For example, the line with gradient 4 crossing the y-axis at -5 has equation $y = 4x - 5$.

> **KEY POINT**
> Straight lines that are **parallel** must have the **same gradient**.

Example Find the equation of the straight line parallel to $y = 3x - 5$ and passing through the point $(4, 2)$.

This equation must be satisfied at the point where $x = 4$ and $y = 2$.

The line must have gradient 3 and so it can be written in the form $y = 3x + c$.
Substituting $x = 4$ and $y = 2$ gives $2 = 3 \times 4 + c \Rightarrow c = -10$.
The required equation is $y = 3x - 10$.

23

Core 1 Pure Mathematics

> **KEY POINT**
> The equation of a straight line with gradient m and passing through the point (x_1, y_1) can be written as $y - y_1 = m(x - x_1)$.

This result is used frequently to find the equation of a tangent or a normal to a curve at a given point.

Using this in the example above gives $y - 2 = 3(x - 4)$. The equation is acceptable in this form but it can be rearranged to give $y = 3x - 10$ as before.

Example Find the equation of the straight line passing through the points $(-1, 5)$ and $(3, -2)$.

One approach is to use the form $y = mx + c$ to produce a pair of simultaneous equations:

Substituting $x = -1$ and $y = 5$ gives $5 = -m + c$ (1)
Substituting $x = 3$ and $y = -2$ gives $-2 = 3m + c$ (2)

(2) − (1) gives: $-7 = 4m \Rightarrow m = \dfrac{-7}{4}$.

Substituting for m in (1) gives: $5 = \dfrac{7}{4} + c \Rightarrow c = \dfrac{13}{4}$.

The equation of the line is $y = \dfrac{-7}{4}x + \dfrac{13}{4}$. This is the same as $4y + 7x = 13$.

An alternative approach is to find the gradient directly and then use the form $y - y_1 = m(x - x_1)$.

Taking $(-1, 5)$ as (x_1, y_1) and $(3, -2)$ as (x_2, y_2),
$m = \dfrac{-2 - 5}{3 - (-1)} = \dfrac{-7}{4} = -\dfrac{7}{4}$.

You can use either of the points $(-1, 5)$ or $(3, -2)$.

The equation of the line is $y - 5 = -\dfrac{7}{4}(x + 1)$.

To show that this is the same as the previous result (you don't *need* to do this):

Multiply both sides of the equation by 4: $4y - 20 = -7(x + 1)$.
Expand the brackets: $4y - 20 = -7x - 7$.
Rearrange to give the previous result: $4y + 7x = 13$.

An equation of the form $ax + by + c = 0$ can be rearranged into gradient–intercept form provided that $b \neq 0$. This becomes $y = -\dfrac{a}{b}x - \dfrac{c}{b}$ and shows that parallel lines may be produced by keeping a and b fixed and allowing c to change.

Note that $4x - 6y = 7$ is equivalent to $2x - 3y = 3.5$

For example, the lines $2x - 3y = 5$, $2x - 3y = -2$, $4x - 6y = 7$ are all parallel.

$m_1 \times m_2 = -1$.

> **KEY POINT**
> When two straight lines are **perpendicular**, the **product** of their **gradients is −1**.

Example Find the equation of the straight line perpendicular to the line $4x + 3y = 12$ and passing through the point $(2, 5)$.

Rearrange the equation into gradient–intercept form: $y = -\dfrac{4}{3}x + 4$.
If the gradient of the required line is m then: $m \times -\dfrac{4}{3} = -1 \Rightarrow m = \dfrac{3}{4}$.
Using the form $y - y_1 = m(x - x_1)$ gives: $y - 5 = \dfrac{3}{4}(x - 2)$.

> **KEY POINT**
>
> If P has coordinates (x_1, y_1) and Q has coordinates (x_2, y_2) then the mid-point of PQ has coordinates $\left(\dfrac{x_1 + x_2}{2}, \dfrac{y_1 + y_2}{2}\right)$.

For example, the mid-point of the line joining $(-3, 1)$ and $(5, 7)$ is $\left(\dfrac{-3 + 5}{2}, \dfrac{1 + 7}{2}\right) = (1, 4)$.

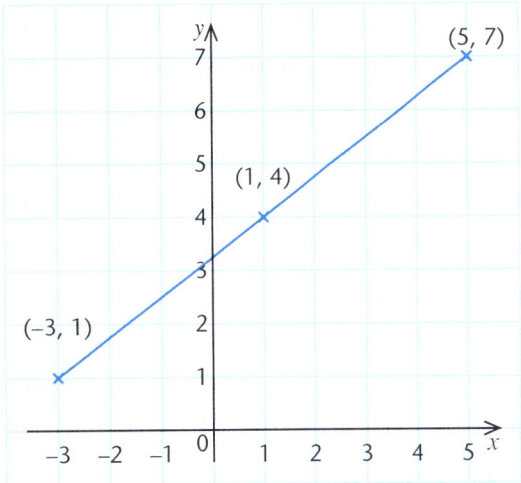

The distance between two points

> **KEY POINT**
>
> The distance between the points $A(x_1, y_1)$ and $B(x_2, y_2)$ is given by $\sqrt{(x_1 - x_2)^2 + (y_1 - y_2)^2}$. The result is based on Pythagoras' theorem.

For example, the distance between $(-3, 4)$ and $(2, -8)$ is given by $\sqrt{(-3 - 2)^2 + (4 - (-8))^2} = \sqrt{5^2 + 12^2} = 13$ units.

The equation of a circle

If the centre is at $(0, 0)$, the equation is $x^2 + y^2 = r^2$

> **KEY POINT**
>
> The equation of a circle with centre (a, b) and radius r is $(x - a)^2 + (y - b)^2 = r^2$.

This result is based on Pythagoras' theorem.
$(x - a)^2 + (y - b)^2 = r^2$

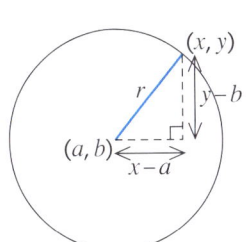

Core 1 Pure Mathematics

For example, the equation of a circle with centre (4, 0) and radius 5 units is $(x-4)^2 + y^2 = 25$.

> **KEY POINT**
>
> An alternative form of the equation of a circle is $x^2 + y^2 + 2gx + 2fy + c = 0$.
> The centre of the circle is $(-g, -f)$ and the radius is $\sqrt{g^2 + f^2 - c}$.

For example, the circle with equation $x^2 + y^2 + 6x - 10y + 18 = 0$ has centre $(-3, 5)$ and radius $\sqrt{(-3)^2 + 5^2 - 18} = \sqrt{16} = 4$ units.

You can also show this by completing the square on the x terms and on the y terms as follows:

$$x^2 + 6x + y^2 - 10y + 18 = 0$$
$$(x+3)^2 - 9 + (y-5)^2 - 25 + 18 = 0$$
$$(x+3)^2 + (y-5)^2 = 16$$

Compare with $(x-a)^2 + (y-b)^2 = r^2$.

The centre is at $(-3, 5)$ and the radius is 4.

Circle properties

You need to remember and be able to use these **circle properties**.

The angle in a **semicircle** is a right angle.

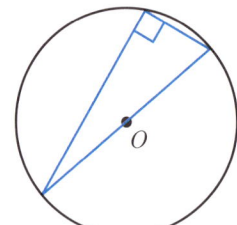

The perpendicular from the centre of a circle to a **chord** bisects the chord.

In the diagram, $AX = XB$.

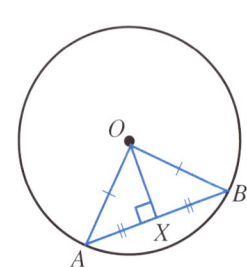

The **tangent** to a circle is perpendicular to the radius at its point of contact.

The normal to the circle at P is along OP, as it is perpendicular to the tangent.

If you know the gradient of OP, you can work out the gradient of the tangent using the fact that for perpendicular lines, the product of the gradients is -1.

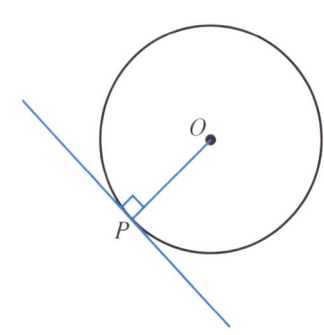

Core 1 Pure Mathematics

Progress check

1. The equation of a straight line is $y = 2x - 7$.
   - (a) Write down the equation of a parallel line crossing the y-axis at 5.
   - (b) Find the equation of a parallel line passing through the point (4, 9).

2. Find the equation of the straight line joining the points (−2, −9) and (1, 3).

3. The equation of a straight line is $3x + 2y = 11$.
   - (a) Find the equation of a parallel line passing through the point (5, 4).
   - (b) Find the equation of a perpendicular line passing through the point (−2, 6).

4. Find the coordinates of the mid-point of (−9, 3) and (−5, 1).

5. P is the point (−1, −4) and Q is the point (5, −1). Find the equation of the line perpendicular to PQ and passing through its mid-point.

6. A is the point (4, −7) and B is the point (−2, 1).
   Find the distance AB.

7. (a) Find the equation of a circle with centre (−3, 8) and radius 5 units.
   (b) A circle has equation $x^2 + y^2 - 12x + 8y + 43 = 0$.
   Find its centre and radius.

> The parallel line can be written in the form $3x + 2y = C$. To find C just substitute the coordinates (5, 4) for x and y.

Answers:
1. (a) $y = 2x + 5$ (b) $y = 2x + 1$
2. $y = 4x - 1$
3. (a) $3x + 2y = 23$ (b) $y - 6 = \frac{2}{3}(x + 2)$ or $3y - 2x - 22 = 0$.
4. (−7, 2)
5. $y = -2x + 1.5$
6. AB = 10 units
7. (a) $(x + 3)^2 + (y - 8)^2 = 25$ (b) (6, −4), 3 units

Core 1 Pure Mathematics

1.3 Differentiation

After studying this section you should be able to:

- use differentiation to find the exact value of the gradient of a curve
- find the equation of the tangent and normal to a curve at a given point
- determine whether a function is increasing or decreasing in an interval
- locate turning points and use the second derivative to distinguish between them

Gradient function

The gradient of a curve changes continuously along its length. Its value at any point P is given by the gradient of the tangent to the curve at P. The gradient may be found approximately by drawing.

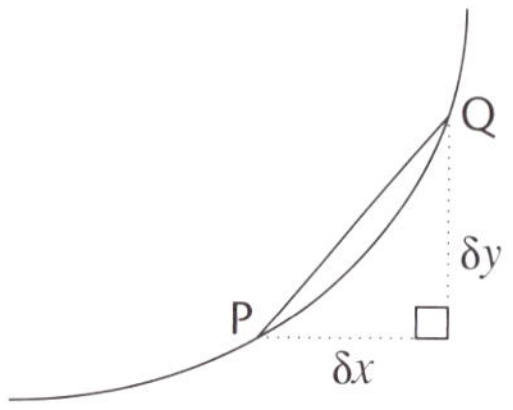

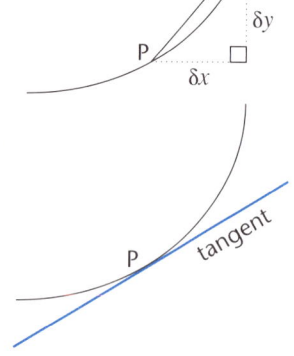

The exact value of the gradient is found by differentiation. This is the limit of the gradient of PQ as Q moves towards P.

Gradient of curve at P $\approx \dfrac{\delta y}{\delta x}$.

As Q moves towards P, $\delta x \rightarrow 0$ and $\dfrac{\delta y}{\delta x} \rightarrow \dfrac{dy}{dx}$.

> **KEY POINT**
>
> $\dfrac{dy}{dx}$ is the gradient function and represents the derivative of y with respect to x.
>
> The graph of $y = kx^n$ has gradient function $\dfrac{dy}{dx} = nkx^{n-1}$.

Example Find the gradient of the curve $y = 3x^2$ at the point P (5, 75).

$\dfrac{dy}{dx} = 2 \times 3x^1 = 6x$.   At P, $x = 5$ so the gradient is $6 \times 5 = 30$.

Function notation may also be used for derivatives. If $y = f(x)$ then $\dfrac{dy}{dx} = f'(x)$. The notation is useful for stating some of the rules of differentiation.

> **KEY POINT**
>
> If $y = f(x) \pm g(x)$ then $\dfrac{dy}{dx} = f'(x) \pm g'(x)$

Example $f(x) = x^3 + 5x + 2$. Find (a) $f'(x)$ (b) $f'(4)$.

Differentiate each term separately. Remember that $x^0 = 1$. When you differentiate a constant, the result is zero.

(a) $f'(x) = 3x^2 + 5$ (b) $f'(4) = 3 \times 4^2 + 5 = 53$.

You often need to express a function in the right form before you can differentiate it.

Example Differentiate: (a) $f(x) = \sqrt{x}$ (b) $f(x) = (x - 5)(x + 3)$ (c) $f(x) = \dfrac{x^3 + 1}{x^2}$.

Core 1 Pure Mathematics

(a) $\sqrt{x}$ has to be written as a power of x to use the rule $f'(x) = nkx^{n-1}$.

$\sqrt{x} = x^{\frac{1}{2}}$ so $f(x) = x^{\frac{1}{2}}$ and $f'(x) = \frac{1}{2}x^{-\frac{1}{2}}$.

> You cannot differentiate a product term by term.

(b) The brackets must first be removed and *then* you can differentiate term by term.

$f(x) = (x-5)(x+3) = x^2 - 2x - 15$.
$f'(x) = 2x - 2$.

> You cannot differentiate a quotient term by term.

(c) Divide $x^3 + 1$ by x^2 first giving $f(x) = x + x^{-2}$.

Now, $f'(x) = 1 - 2x^{-3} = 1 - \dfrac{2}{x^3}$.

Tangents and normals

You can find the gradient of the tangent to a curve at a point by differentiation. Then you can use the techniques described in the Coordinate Geometry section to find the **equation of the tangent** and the **equation of the normal** at the given point.

Example Find the equation of the tangent and the normal to the curve $y = x^3 - 4x$ at the point (2, 0).

Differentiate the equation of the curve to give $\dfrac{dy}{dx} = 3x^2 - 4$.

The gradient of the tangent at (2, 0) is $3 \times 2^2 - 4 = 8$.
Using $y - y_1 = m(x - x_1)$ gives the equation of the tangent as $y = 8(x - 2)$.

> The normal at a point is perpendicular to the tangent.

The gradient of the normal is $-\frac{1}{8}$. Using $y - y_1 = m(x - x_1)$ again gives the equation of the normal as $y = -\frac{1}{8}(x - 2)$. You could rearrange this to give $x + 8y = 2$.

Curve sketching and stationary points

The gradient function gives information about the behaviour of a curve.

$f'(x) > 0$ at A and the function is **increasing**.

$f'(x) < 0$ at C and the function is **decreasing**.

$f'(x) = 0$ at B and D and the function is neither increasing nor decreasing. These points are called **stationary points**. B is at a **local maximum** and D is at a **local minimum**.

> You are not required to identify points of inflexion.

E and F show a different type of stationary point, called a **stationary point of inflexion**.

One way to distinguish between the three types of stationary point is to look at the sign of the derivative on either side of the point.

> The first two types of stationary point are also known as **turning points**.

- At a local maximum the sign of the derivative changes from positive to negative.
- At a local minimum the sign of the derivative changes from negative to positive.
- At a stationary point of inflexion there is no change of sign.

The second derivative

Starting from $\dfrac{dy}{dx} = f'(x)$ and differentiating again gives the **second derivative**. This is written as $\dfrac{d^2y}{dx^2} = f''(x)$. The second derivative can give information about the nature of any stationary points.

At a stationary point:

- $f''(x) > 0 \Rightarrow$ the point is a local minimum
- $f''(x) < 0 \Rightarrow$ the point is a local maximum.

But, if $f''(x) = 0$ then this gives no further information.

Example Find the stationary points of the curve $y = \dfrac{x^3}{3} - \dfrac{3}{2}x^2 + 2x - 1$ and use the second derivative to distinguish between them.

First differentiate the equation, then solve $\dfrac{dy}{dx} = 0$ to locate the stationary points.

$\dfrac{dy}{dx} = x^2 - 3x + 2.$

At a stationary point $\dfrac{dy}{dx} = 0 \Rightarrow x^2 - 3x + 2 = 0$

$\Rightarrow (x-1)(x-2) = 0$

$\Rightarrow x = 1 \text{ or } x = 2.$

When $x = 1$, $y = \tfrac{1}{3} - \tfrac{3}{2} + 2 - 1 = -\tfrac{1}{6}$.

When $x = 2$, $y = \tfrac{8}{3} - 6 + 4 - 1 = -\tfrac{1}{3}$.

The stationary points are $(1, -\tfrac{1}{6})$ and $(2, -\tfrac{1}{3})$.

In this case, the function is a cubic and so the shape of the curve is known. We should expect the first stationary point to be a local maximum and the second to be a local minimum.

Using the second derivative: $f''(x) = 2x - 3$

so $f''(1) = -1 < 0$ giving a local maximum,

and $f''(2) = 1 > 0$ giving a local minimum as expected.

Note:

> **KEY POINT**
> If $y = kx^n$, then $\dfrac{dy}{dx} = nkx^{n-1}$ for all rational values of n.

You also have to differentiate expressions when n is not necessarily a positive integer. You may have to change the expression into index form first.

Example 1 Differentiate $y = x^{\tfrac{3}{2}} + \dfrac{3}{x^2}$

Re-write as $y = x^{\tfrac{3}{2}} + 3x^{-2}$ and differentiate term by term to give

$\dfrac{dy}{dx} = \dfrac{3}{2}x^{\tfrac{1}{2}} + (-2) \times 3x^{-3}$

$= \dfrac{3}{2}\sqrt{x} - \dfrac{6}{x^3}$

Example 2 Differentiate $y = x\sqrt{x}$

Re-write as $y = x^{\tfrac{3}{2}}$ giving

$\dfrac{dy}{dx} = \dfrac{3}{2}\sqrt{x}$

Progress check

1 Differentiate with respect to x:
$y = 6\sqrt{x}$.

2 Differentiate with respect to x:
$f(x) = \dfrac{2x+1}{x^3}$.

3 Find the gradient of the curve $y = (4x - 1)^2$ at the point $(1, 9)$.

> Solve the inequality $f'(x) < 0$.

4 Find the values of x for which the function $f(x) = x^3 - 6x^2 + 9x - 7$ is decreasing.

5 Find the equation of: (a) the tangent (b) the normal
to the curve $y = x^3 - 2x$ at the point $(2, 4)$.

6 Find $\dfrac{dy}{dx}$ in each of the following questions.

(a) $y = x^{\frac{5}{3}} - \dfrac{1}{x}$

(b) $y = x^3\sqrt{x}$

---

1 $\dfrac{3}{\sqrt{x}}$

2 $\dfrac{4}{x^3} - \dfrac{3}{x^4}$

3 24

4 $1 < x < 3$

5 (a) Tangent: $y = 10x - 16$; (b) normal: $x + 10y - 42 = 0$.

6 (a) $\dfrac{5}{3}x^{\frac{2}{3}} + \dfrac{1}{x^2}$

(b) $\dfrac{7}{2}x^{\frac{5}{2}}$ or $\dfrac{7}{2}x^2\sqrt{x}$ or $\dfrac{7}{2}x^3\sqrt{x}$

Core 1 Pure Mathematics

Sample questions and model answers

'Write down' means that you don't need to show any working to get the marks. But you might still find it useful as a way to clarify your thoughts.

Remember that $a^{-n} = \dfrac{1}{a^n}$

1 Write down the exact value of 3^{-3}.

$3^{-3} = \dfrac{1}{3^3} = \dfrac{1}{27}$.

2

(a) Write $2x^2 - 12x + 11$ in the form $a(x+b)^2 + c$.

(b) State the minimum value of $2x^2 - 12x + 11$ and give the value of x where this occurs.

(c) Solve the equation $2x^2 - 12x + 11 = 0$ and express your answer in surd form.

(d) Sketch the graph of $y = 2x^2 - 12x + 11$.

Start by taking out the factor of 2 common to the first two terms.

Keep the 11 separate rather than have a fraction inside the brackets.

Complete the square and simplify.

(a) $2x^2 - 12x + 11 = 2(x^2 - 6x) + 11$

$= 2((x-3)^2 - 9) + 11$

$= 2(x-3)^2 - 7$.

The minimum value occurs when the expression in the brackets equals zero.

(b) The minimum value is -7 and this occurs when $x = 3$.

(c) $2x^2 - 12x + 11 = 0$

$\Rightarrow 2(x-3)^2 - 7 = 0$

$\Rightarrow (x-3)^2 = \dfrac{7}{2} = \dfrac{14}{4}$

$\Rightarrow x - 3 = \pm \dfrac{1}{2}\sqrt{14}$

$\Rightarrow x = 3 \pm \dfrac{1}{2}\sqrt{14}$.

(d)

When $x = 0$, $y = 11$.

Use the information you have found to position the curve and label the key points.

32

Sample questions and model answers (continued)

3

A curve has equation $y = (2 - x)(2 + x)$.

The point P(−1, 3) lies on the curve.

(a) Find the gradient of the tangent to the curve at P.

(b) Find the equation of the normal to the curve at P in the form $px + qy = r$.

(c) The normal to the curve at P crosses the y-axis at A and the x-axis at B. Find the area of triangle AOB.

> Expand the brackets before differentiating.

(a) $y = (2 - x)(2 + x)$
$ = 4 - x^2$
$\dfrac{dy}{dx} = -2x.$

When $x = -1$, $\dfrac{dy}{dx} = -2(-1) = 2$

The gradient of the tangent at P is 2.

(b) The gradient of the normal at P is $-\tfrac{1}{2}$.

> The product of the gradients must be −1, as they are at right angles to each other.
>
> This is using the equation of a straight line in the form $y - y_1 = m(x - x_1)$.
>
> Write the equation in the required form.

Equation of the normal at P:
$y - 3 = -\tfrac{1}{2}(x - (-1))$
$y - 3 = -\tfrac{1}{2}(x + 1)$
$2(y - 3) = -(x + 1)$
$2y - 6 = -x - 1$
$x + 2y = 5$

> Find where the line intersects the axes.

(c) When $x = 0$, $y = \tfrac{5}{2}$ ⇒ A is the point $(0, \tfrac{5}{2})$
When $y = 0$, $x = 5$ ⇒ B is the point $(5, 0)$

Area of triangle AOB $= \tfrac{1}{2} \times 5 \times \tfrac{5}{2} = \tfrac{25}{4} = 6\tfrac{1}{4}$ units^2.

Core 1 Pure Mathematics

Practice examination questions

1. Write down the exact value of $49^{-\frac{1}{2}}$.

2. Given that $8^{x-3} = 4^{x+1}$, find the value of x.

3. (a) Write $x^2 - 4x + 1$ in the form $(x + a)^2 + b$.

   (b) Solve $x^2 - 4x + 1 = 0$ and express your answer in surd form.

   (c) State the coordinates of the lowest point on the graph of $y = x^2 - 4x + 1$.

   (d) Sketch the graph.

4. Given that $f(x) = x^2 + \dfrac{1}{x}$, find

   (a) $f'(2)$.

   (b) $f''(-1)$.

5. Find the coordinates of the points A and B at the intersection of these graphs.

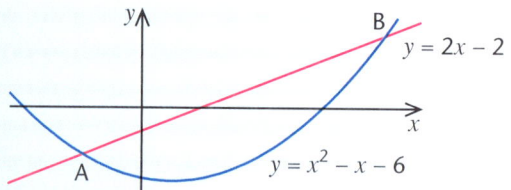

6. Solve the inequality

   $3x^2 - 8x - 7 < 2x^2 - 3x - 11$.

7. (a) The diagram shows a straight line passing through the points A(−1, −5) and B(4, 1).
   Find its equation in the form $ax + by + c = 0$.

   (b) A second line l, passes through the mid-point of A and B at right angles to AB.
   Find its equation in the form $ax + by + c = 0$.

   (c) The line l crosses the y-axis at C. Find the coordinates of C.

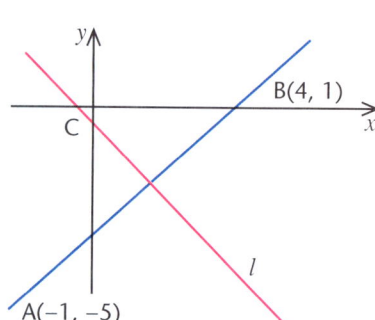

Practice examination questions (continued)

8 The diagram shows two lines l and m at right angles to each other. The equation of the line m is $3x + 2y = 6$. The line l passes through the point with coordinates (5, 4). l and m cross the x-axis at A and B respectively.

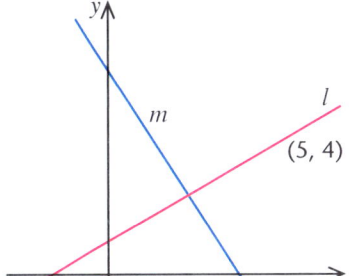

(a) Find the equation of the line l.

(b) Show that AB = 3.

9 (a) Find the coordinates of the stationary points on the curve $y = 2x^3 - 15x^2 - 36x + 10$.

(b) Use the second derivative to determine the nature of the stationary points found in (a).

(c) Find the set of values of x for which $2x^3 - 15x^2 - 36x + 10$ is a decreasing function of x.

10 The diagram shows a straight line crossing the curve $y = (x - 1)^2 + 4$ at the points P and Q. It passes through (0, 7) and (7, 0).

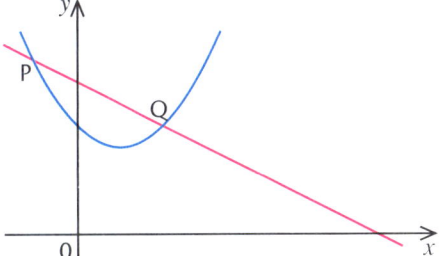

(a) Write down the equation of the straight line through P and Q.

(b) Find the coordinates of P and Q.

11 (a) Factorise $2x^{\frac{3}{2}} - \dfrac{2x^{\frac{5}{2}}}{5}$.

(b) Find the points at which the graph of $y = 2x^{\frac{3}{2}} - \dfrac{2x^{\frac{5}{2}}}{5}$ meets the x-axis.

(c) Find $\dfrac{dy}{dx}$.

(d) Solve $\dfrac{dy}{dx} = 0$ for $x > 0$. Hence find the coordinates of the turning point of the graph, where x is positive, and determine its nature.

Chapter 2
Core 2 Pure Mathematics

The following topics are covered in this chapter:

- Algebra and functions
- Sequences and series
- Trigonometry
- Integration

2.1 Algebra and functions

After studying this section you should be able to:

- understand exponential functions and sketch their graphs
- understand the log laws and use them to solve equations of the form $a^x = b$
- carry out algebraic division
- use the remainder and factor theorems

Exponential functions

An **exponential function** is one where the variable is a power or exponent. For example, any function of the form $f(x) = a^x$, where a is a constant, is an exponential function.

The diagram shows the graphs of $y = 2^x$ and $y = 10^x$.

All graphs of the form $y = a^x$ cross the y-axis at (0, 1). Each graph has a different gradient at this point and the value of the gradient depends on a.

Exponential functions are often used to represent, or model, patterns of:
- growth when $a > 1$
- decay when $a < 1$.

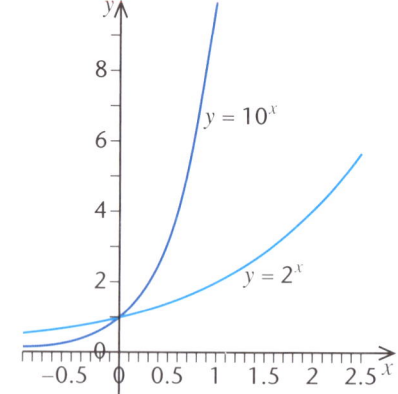

Logarithms

Exponential functions are related to **logarithmic functions** as follows:

$$y = \log_a x \iff x = a^y$$

$a > 0$ and $a \neq 1$.

a is called the base of the logarithm.

Your calculator has a key for logs to the base 10, probably labelled log.

The **logarithm** (log) to a given **base** of a number is the **power** to which the base must be raised to equal the number.

For example, $10^2 = 100$, so $\log_{10} 100 = 2$.

It is useful to remember the following hold for any base a:

$\log_a a = 1$, since $a^1 = a$

$\log_a 1 = 0$, since $a^0 = 1$.

The graph of $y = \log_a x$ is a reflection of $y = a^x$ in the line $y = x$.

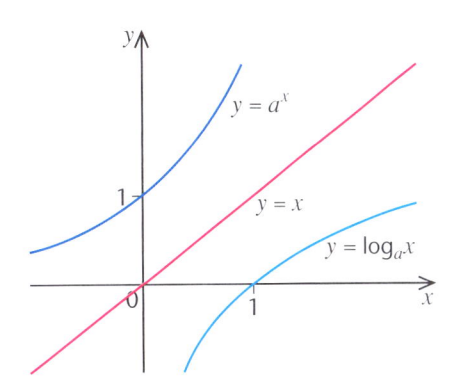

The log laws

There are three laws of logarithms that you need to know. The same laws apply in any base and so no particular base is stated:

- $\log a + \log b = \log ab$
- $\log a - \log b = \log\left(\dfrac{a}{b}\right)$
- $n \log a = \log a^n$.

You can use these laws to simplify expressions involving logs and to solve **exponential equations**, i.e. equations where the unknown value is a power.

Example Express $\log x + 3 \log y$ as a single logarithm.

$$\log x + 3 \log y = \log x + \log y^3 \quad \text{(using the third law)}$$
$$= \log xy^3 \quad \text{(using the first law)}$$

Example Solve the equation $5^x = 30$.

Taking logarithms of both sides gives:

$$\log 5^x = \log 30$$
$$\Rightarrow x \log 5 = \log 30$$
$$\Rightarrow x = \frac{\log 30}{\log 5} = 2.113 \text{ to 4 s.f.}$$

This works in any base.

Algebraic division

If you work out $27 \div 4$ for example, then you obtain 6 as the **quotient** and 3 as the **remainder**. You can write $27 = 4 \times 6 + 3$ to show the connection between these values.

The same thing applies in algebra when one polynomial is divided by another.

Example Find the quotient and remainder when $x^2 + 7x - 5$ is divided by $x - 4$.

Using $ax + b$ for the quotient and c for the remainder gives:

$$x^2 + 7x - 5 \equiv (x - 4)(ax + b) + c.$$
$$\equiv ax^2 + (b - 4a)x + c - 4b.$$

Core 2 Pure Mathematics

Equating coefficients of x^2 gives: $a = 1$.

Equating coefficients of x gives: $b - 4a = 7$

$\Rightarrow b - 4 = 7$ (using the result $a = 1$ from above)

$\Rightarrow b = 11$.

Equating the constant terms gives: $c - 4b = -5$

$\Rightarrow c - 44 = -5$ (using the result $b = 11$ from above)

$\Rightarrow c = 39$.

Substituting for a, b and c in [1] gives $x^2 + 7x - 5 \equiv (x - 4)(x + 11) + 39$.

The remainder theorem

In the previous example, the quotient and remainder of $(x^2 + 7x - 5) \div (x - 4)$ were found by using $x^2 + 7x - 5 \equiv (x - 4)(ax + b) + c$ and equating coefficients.

When $x = 4$, $(x - 4) = 0$ so $(x - 4)(ax + b) = 0$ and the RHS simplifies to c.

Another way to use this identity is to substitute particular values for x. Notice that when $x = 4$, the RHS simplifies to c and so the remainder is easily found by substituting $x = 4$ on the LHS. This gives $c = 4^2 + 7 \times 4 - 5 = 39$ as before.

In its generalised form this result is known as the **remainder theorem**, which states:

> When a polynomial $f(x)$ is divided by $(x - a)$ the remainder is $f(a)$.
>
> **KEY POINT**

Example Find the remainder when the polynomial $f(x) = x^3 + x - 5$ is divided by:

(a) $(x - 3)$ (b) $(x + 2)$ (c) $(2x - 1)$.

Choose x so that $x - 3 = 0$.

(a) The remainder is $f(3) = 3^3 + 3 - 5 = 25$

(b) The remainder is $f(-2) = (-2)^3 - 2 - 5 = -15$

Choose x so that $2x - 1 = 0$.

(c) The remainder is $f(0.5) = 0.5^3 + 0.5 - 5 = -4.375$

The factor theorem

A special case of the remainder theorem occurs when the remainder is 0. This result is known as the **factor theorem**, which states:

> If $f(x)$ is a polynomial and $f(a) = 0$ then $(x - a)$ is a factor of $f(x)$.
>
> **KEY POINT**

Example Show that $(x + 3)$ is a factor of $x^3 + 5x^2 + 5x - 3$.

Taking $f(x) = x^3 + 5x^2 + 5x - 3$

$f(-3) = (-3)^3 + 5(-3)^2 + 5(-3) - 3$

$= -27 + 45 - 15 - 3 = 0$.

By the factor theorem $(x + 3)$ is a factor of $x^3 + 5x^2 + 5x - 3$.

Core 2 Pure Mathematics

Example A polynomial is given by $f(x) = 2x^3 + 13x^2 + 13x - 10$.

(a) Find the value of $f(2)$ and $f(-2)$.

(b) State one of the factors of $f(x)$.

(c) Factorise $f(x)$ completely.

(a) $f(2) = 2(2^3) + 13(2^2) + 13(2) - 10 = 84$

$f(-2) = 2(-2)^3 + 13(-2)^2 + 13(-2) - 10 = 0.$

(b) Since $f(-2) = 0$, $(x+2)$ is a factor of $f(x)$ by the factor theorem.

$f(x)$ is a cubic so the linear factor $(x+2)$ must be combined with a quadratic.

(c) $2x^3 + 13x^2 + 13x - 10 \equiv (x+2)(ax^2 + bx + c)$.

This is an **identity** and so the values of a, b and c may be found by comparing coefficients.

Comparing the x^3 terms gives:	$a = 2.$
Comparing the x^2 terms gives:	$13 = 2a + b \Rightarrow b = 9.$
Comparing the constant terms gives:	$c = -5.$
It follows that:	$f(x) = (x+2)(2x^2 + 9x - 5).$

Factorising the quadratic part in the usual way gives $f(x) = (x+2)(2x-1)(x+5)$.

Progress check

1. Solve the equation $4^x = 100$ and give your answer to 4 s.f.

2. Write each of the following as a single logarithm.
   (a) $\log_a x + 3\log_a y - \tfrac{1}{2}\log_a z$
   (b) $\log_{10} x - 1$

3. Evaluate the following:
   (a) $\log_2 8$
   (b) $\log_9 3$
   (c) $2\log_4 2$

4. Find the remainder when $x^3 + 2x^2 - 5x + 1$ is divided by $(x-2)$.

5. Show that $x+7$ is a factor of $x^3 - x^2 - 65x - 64$ and hence factorise the expression completely.

6. Show that $(x+1)$ is a factor of $x^3 + 11x^2 + 31x + 21$ and hence solve the equation $x^3 + 11x^2 + 31x + 21 = 0$.

1 $x = 3.322$
2 (a) $\log_a\left(\dfrac{xy^3}{\sqrt{z}}\right)$ (b) $\log_{10}\left(\dfrac{x}{10}\right)$
3 (a) 3 (b) $\tfrac{1}{2}$ (c) 1
4 7
5 $(x+7)(x+1)(x-9)$
6 $-1, -3, -7$

2.2 Sequences and series

After studying this section you should be able to:

- understand the different types of sequence, including the notation and formulae used to describe them
- recognise arithmetic and geometric progressions and calculate the sum of their series
- determine whether a geometric series with an infinite number of terms has a finite sum and calculate its value when it exists

Sequences and series

A list of numbers in a particular order, that follow some rule for finding later values, is called a **sequence**. Each number in a sequence is called a **term**, and terms are often denoted by $u_1, u_2, u_3, \ldots, u_n, \ldots$.

One way to define a sequence is to give a formula for the nth term such as $u_n = n^2$. Substituting the values $n = 1, 2, 3, 4, \ldots$ produces the sequence $1, 4, 9, 16, \ldots$ and the value of any particular term can be calculated by substituting its position number into the formula. For example, in this case, the 50th term is $u_{50} = 50^2 = 2500$.

Another way to define a sequence is to give a starting value together with a rule that shows the connection between successive terms. This is sometimes called a **recursive definition**. For example $u_1 = 5$ and $u_{n+1} = 2u_n$ defines the sequence $5, 10, 20, 40, \ldots$. The rule $u_{n+1} = 2u_n$ is an example of a **recurrence relation**.

Two special sequences are the **arithmetic progression** (A.P.) and the **geometric progression** (G.P.).

In an A.P. successive terms have a **common difference**, e.g. $1, 4, 7, 10, \ldots$. The first term is denoted by a and the common difference is d. With this notation, the definition of an A.P. may now be given as $u_1 = a$, $u_{n+1} = u_n + d$.

The terms of an A.P. take the form $a, a+d, a+2d, a+3d, \ldots$ and the nth term is given by $u_n = a + (n-1)d$.

In a G.P. successive terms are connected by a **common ratio**, e.g. $3, 6, 12, 24, \ldots$. The definition of a G.P. may be given as $u_1 = a$, $u_{n+1} = ru_n$, where r is the common ratio.

The terms of a G.P. take the form $a, ar, ar^2, ar^3, \ldots$ and the nth term is given by $u_n = ar^{n-1}$.

A **series** is formed by adding together the terms of a sequence. The use of sigma notation can greatly simplify the way that series are written. For example, the series $1^2 + 2^2 + 3^2 + \ldots + n^2$ may be written as $\sum_{i=1}^{n} i^2$. The sum of the first n terms of a series is often denoted by S_n and so $S_n = u_1 + u_2 + u_3 + \ldots + u_n = \sum_{i=1}^{n} u_i$.

Core 2 Pure Mathematics

Arithmetic series

The **sum of an arithmetic series** is given by
$$S_n = a + (a + d) + (a + 2d) + \ldots + (a + (n-1)d).$$

This may also be written as
$$S_n = l + (l - d) + (l - 2d) + \ldots + (l - (n-1)d),$$
where l is the last term.

Adding the two versions gives
$$2S_n = (a + l) + (a + l) + (a + l) + \ldots (a + l)$$
$$= n(a + l).$$
$$S_n = \frac{n}{2}(a + l).$$

You need to know how to establish the general result.

> **KEY POINT**
>
> The sum of the first n terms of an arithmetic series is: $S_n = \frac{n}{2}(a + l)$.
>
> Substituting $l = a + (n-1)d$ gives the alternative form of the result:
> $$S_n = \frac{n}{2}(2a + (n-1)d).$$

A special case is the sum of the first n natural numbers:
$$1 + 2 + 3 + \ldots + n = \frac{n}{2}(n + 1).$$

For example, the sum of the first 100 natural numbers is $\frac{100}{2}(100 + 1) = 5050$.

When finding the sum of an A.P. you need to select the most appropriate version of the formula to suit the information.

Example Find the sum of the first 50 terms of the series $15 + 18 + 21 + 24 + \ldots$.

In this series, $a = 15$, $d = 3$ and $n = 50$.

Using $S_n = \frac{n}{2}(2a + (n-1)d)$ gives $S_{50} = \frac{50}{2}(2 \times 15 + 49 \times 3) = 4425$.

Geometric series

The **sum of a G.P.** is given by $S_n = a + ar + ar^2 + ar^3 + \ldots + ar^{n-1}$.

Multiplying throughout by r gives $rS_n = ar + ar^2 + ar^3 + \ldots + ar^{n-1} + ar^n$.

Subtracting gives:
$$S_n - rS_n = a - ar^n$$
$$\Rightarrow S_n(1 - r) = a(1 - r^n)$$

You need to know how to establish this result.

> **KEY POINT**
>
> So the sum of the first n terms of a geometric series is $S_n = \frac{a(1 - r^n)}{1 - r}$.

If $r > 1$ then it is more convenient to use the result in the form $S_n = \frac{a(r^n - 1)}{r - 1}$.

Example Find the sum of the first 20 terms of the series $8 + 12 + 18 + 27 + \ldots$ to the nearest whole number.

In this series, $a = 8$, $r = 1.5$ and $n = 20$.

This gives $S_{20} = \frac{8(1.5^{20} - 1)}{1.5 - 1} = 53188.107\ldots = 53188$ to the nearest whole number.

Core 2 Pure Mathematics

> Provided that $|r| < 1$, the sum of a geometric series converges to $\dfrac{a}{1-r}$ as n tends to infinity. This is known as the **sum to infinity** of a geometric series.

KEY POINT

For example, $1 + \dfrac{1}{2} + \dfrac{1}{4} + \dfrac{1}{8} + \ldots = \dfrac{1}{1 - \frac{1}{2}} = 2$.

Binomial expansion

The situation where n is not a positive integer can be dealt with but you don't need it for this component.

An expression which has two terms, such as $a + b$ is called a **binomial**. The expansion of something of the form $(a+b)^n$ is called a **binomial expansion**. When n is a positive integer:

$$(a+b)^n = a^n + na^{n-1}b + \frac{n(n-1)}{2!}a^{n-2}b^2 + \frac{n(n-1)(n-2)}{3!}a^{n-3}b^3 + \ldots + b^n.$$

This expansion may appear complicated but it starts to look simpler if you follow the patterns from one term to the next:

Remember that $a^0 = 1$ when $a \neq 0$.

- Starting with a^n, the power of a is reduced by 1 each time until the last term, b^n, which is the same as $a^0 b^n$.
- The power of b is increased by 1 each time. Notice that the first term, a^n, is the same as $a^n b^0$ and that the powers of a and b always add up to n in each term.
- It's worth remembering the first three coefficients: 1, n and $\dfrac{n(n-1)}{2!}$ then the rest follow the same pattern, giving $\dfrac{n(n-1)(n-2)}{3!}$, $\dfrac{n(n-1)(n-2)(n-3)}{4!}$ and so on.

Pascal's triangle: each row starts and ends with 1. Every other value is found by adding the pair of numbers immediately above it in the pattern.

```
       1   4   6
   1   5   10
```

The coefficients in the expansion of $(a+b)^n$ also follow the pattern given by the row of **Pascal's triangle** that starts with $1 \quad n \quad \ldots$
So the coefficients in the expansion of $(a+b)^4$ for example are $1, 4, 6, 4$ and 1.
It follows that
$(a+b)^4 = a^4 + 4a^3 b + 6a^2 b^2 + 4ab^3 + b^4$.

```
              1
            1   1
          1   2   1
        1   3   3   1
      1   4   6   4   1
    1   5  10  10   5   1
```

Variations on this result can be obtained by substituting different values for a and b. Some **examples** are:

In this case, $a = 1$ and $b = x$.

$$(1+x)^4 = 1 + 4x + 6x^2 + 4x^3 + x^4.$$

and

For this one, $a = 1$ and $b = -2x$.

$$(1-2x)^4 = 1 + 4(-2x) + 6(-2x)^2 + 4(-2x)^3 + (-2x)^4$$
$$= 1 - 8x + 24x^2 - 32x^3 + 16x^4.$$

The notation $\binom{n}{r}$ is often used to stand for the expression $\dfrac{n(n-1)\ldots(n-r+1)}{r!}$.

So, for example $\binom{n}{3} = \dfrac{n(n-1)(n-2)}{3!}$.

Using this notation, the binomial expansion may be written as:

$$(a+b)^n = a^n + \binom{n}{1}a^{n-1}b + \binom{n}{2}a^{n-2}b^2 + \binom{n}{3}a^{n-3}b^3 + \ldots + b^n.$$

- It's useful to recognise that the term involving b^r takes the form $\binom{n}{r}a^{n-r}b^r$.

- For positive integer values $\binom{n}{r}$ has the same value as ^nC_r and you may find that your calculator will work this out for you.

- When n is small and the full binomial expansion is required, the simplest way to find the coefficients is to use Pascal's triangle. For larger values of n it may be simpler to use the formula, particularly if only some of the terms are required.

Example Expand $(1 + 3x)^{10}$ in ascending powers of x up to and including the fourth term.

$$(1+3x)^{10} = 1 + \binom{10}{1}(3x) + \binom{10}{2}(3x)^2 + \binom{10}{3}(3x)^3 + \ldots$$

$$= 1 + 10(3x) + 45(9x^2) + 120(27x^3) + \ldots$$

$$= 1 + 30x + 405x^2 + 3240x^3 + \ldots.$$

Example Find the coefficient of the x^7 term in the expansion of $(3 - 2x)^{15}$

The term involving x^7 is given by $\binom{15}{7}(3)^8(-2x)^7$

$$= 6435 \times 6561 \times -128x^7,$$

and so the required coefficient is $-5\,404\,164\,480$.

Notice that the coefficient of x^7 is not totally determined by the value of $\binom{15}{7}$.

Progress check

1 Write down the first five terms of the sequence given by:
   (a) $u_n = 2n^2$ (b) $u_1 = 10$, $u_{n+1} = 3u_n + 2$.

2 Find the 20th term of an A.P. with first term 7 and common difference 5.

3 Find the sum of the first 1000 natural numbers.

4 Find the sum of the first 20 terms of $12 + 15 + 18.75 + \ldots$ to 4 s.f.

5 Explain why the series $10 + 9 + 8.1 + 7.29 + \ldots$ is convergent and find the value of its sum to infinity to 5 s.f.

1 (a) 2, 8, 18, 32, 50 (b) 10, 32, 98, 296, 890
2 102
3 500 500
4 4115
5 $|r| = 0.9 < 1$, 100.

Core 2 Pure Mathematics

2.3 Trigonometry

After studying this section you should be able to:

- state the exact values of the sine, cosine and tangent of special angles
- understand the properties of the sine, cosine and tangent functions
- sketch the graphs of trigonometric functions
- use the sine and cosine rules and the formula for the area of a triangle
- understand radian measure and use it to find arc lengths and areas of sectors
- solve trigonometric equations in a given interval

Sine, cosine and tangent of special angles

The sine, cosine and tangent of 30°, 45°, and 60° may be expressed exactly as shown below. The results are based on Pythagoras' theorem.

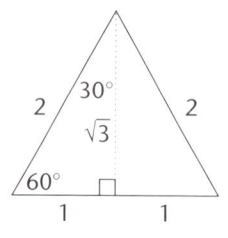

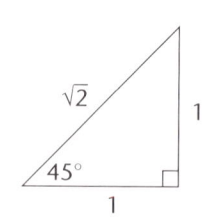

$$\sin 30° = \frac{1}{2} \qquad \sin 60° = \frac{\sqrt{3}}{2} \qquad \sin 45° = \frac{1}{\sqrt{2}}$$

$$\cos 30° = \frac{\sqrt{3}}{2} \qquad \cos 60° = \frac{1}{2} \qquad \cos 45° = \frac{1}{\sqrt{2}}$$

$$\tan 30° = \frac{1}{\sqrt{3}} \qquad \tan 60° = \sqrt{3} \qquad \tan 45° = 1.$$

Trigonometric graphs

You need to be able to sketch the graphs of the sine, cosine and tangent functions and to know their special properties.

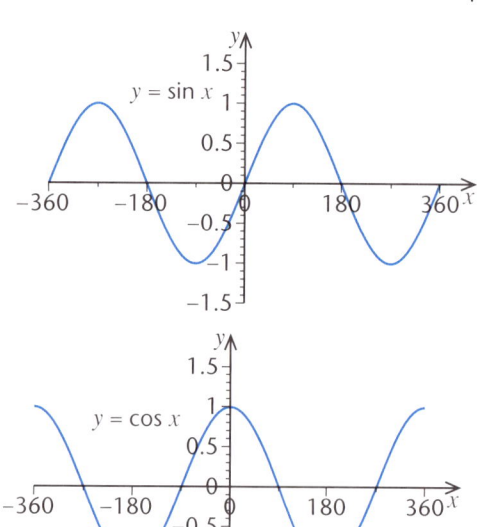

$\sin x$ is defined for any angle and always has a value between -1 and 1. It is a **periodic function** with period $360°$.

The graph has **rotational symmetry** of order 2 about every point where it crosses the x-axis.

It has **line symmetry** about every vertical line passing through a vertex.

$\cos x \equiv \sin(x + 90°)$ so the graph of $y = \cos x$ can be obtained by translating the sine graph $90°$ to the left.

It follows that $\cos x$ is also a periodic function with period $360°$ and has the corresponding symmetry properties.

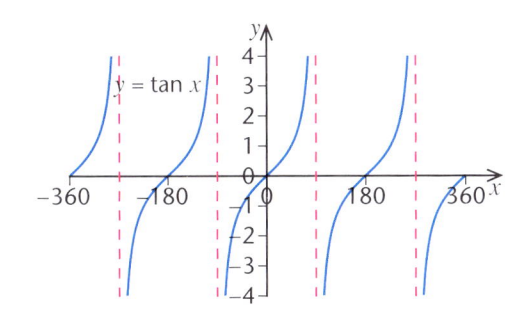

Core 2 Pure Mathematics

$$\tan x \equiv \frac{\sin x}{\cos x}.$$

$\tan x$ is undefined whenever $\cos x = 0$ and approaches $\pm\infty$ near these values. It is a periodic function with period $180°$.

The graph has rotational symmetry of order 2 about $0°$, $\pm 90°$, $\pm 180°$, $\pm 270°$, ….

The graphs of more complex trigonometric functions can often be produced by applying transformations to one of the basic graphs.

Example Sketch the graph of $y = 2\sin(x + 30°) + 1$ for $0° \leqslant x \leqslant 360°$.

It is helpful to think about building the transformations in stages.

		Transformations		
Basic function	Translate the curve $30°$ to the left.		Now apply a one-way stretch with scale factor 2 parallel to the y-axis.	Finally translate the curve 1 unit up.
$y = \sin x$	$y = \sin(x + 30)$		$y = 2\sin(x + 30)$	$y = 2\sin(x + 30°) + 1$

The graph of $y = 2\sin(x + 30°) + 1$ may now be produced by applying these transformations in order starting from the graph of $y = \sin x$.

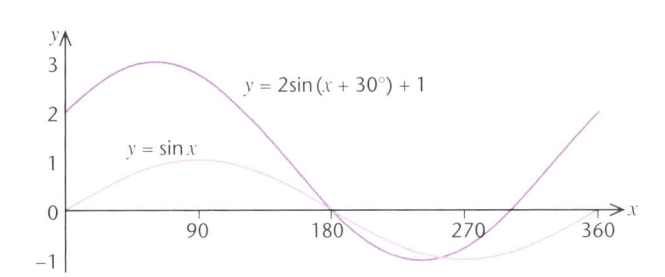

Area of a triangle

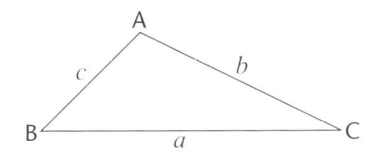

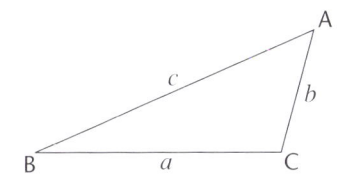

Area of triangle $= \frac{1}{2}ab \sin C$

In both the diagrams, $b \sin C$ represents the height of the triangle.

The same formula applies whether the angle is acute or obtuse.

The sine and cosine rules

A, B and C can be at any vertex. The opposite sides are then labelled as a, b, c.

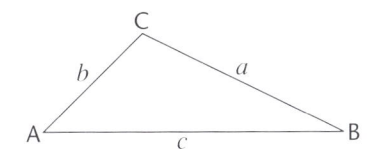

Sine rule: $\dfrac{a}{\sin A} = \dfrac{b}{\sin B} = \dfrac{c}{\sin C}$

Cosine rule: $a^2 = b^2 + c^2 - 2bc \cos A$

This may be rearranged to find an angle

$$\cos A = \frac{b^2 + c^2 - a^2}{2bc}$$

45

Core 2 Pure Mathematics

Radians

Angles can also be measured in **radians** and this makes it much easier to deal with trigonometric functions when using **calculus**.

1 radian ≈ 57.3°. This may be written as 1^c ≈ 57.3°. However, the symbol for radians is not normally written when the angle involves π.
The following results are useful to remember:

$\pi = 180°, \dfrac{\pi}{2} = 90°, \dfrac{\pi}{4} = 45°, \dfrac{\pi}{3} = 60°, \dfrac{\pi}{6} = 30°$.

All the results that you know in degrees also apply in radians, e.g. $\sin\dfrac{\pi}{6} = 0.5$

Arc length and sector area

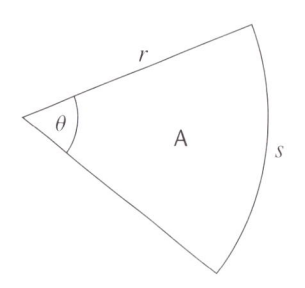

With θ in radians:

- Length of arc of sector is given by
  $s = r\theta$.

- Area of sector is given by
  $A = \tfrac{1}{2}r^2\theta$.

Solving trigonometric equations

You may be required to solve trigonometric equations in a variety of forms.

Example Solve $\sin(2x - 30°) = 0.6$ for $0° \leqslant x \leqslant 360°$.

One value of $2x - 30°$ is $\sin^{-1}(0.6) = 36.86...°$. *This is the principal value given by the calculator.*

First find a value of $2x - 30°$.

You need to adjust the interval to find where the values of $2x - 30°$ must lie.

$0° \leqslant x \leqslant 360° \Rightarrow 0° \leqslant 2x \leqslant 720° \Rightarrow -30° \leqslant 2x - 30° \leqslant 690°$.

Now look at the graph of the sine function in this interval.

This is the graph of $y = \sin X$ for $-30° \leqslant X \leqslant 690°$ where $X = 2x - 30°$. In other words, draw the sine curve as normal for the interval and then interpret the variable as $2x - 30°$.

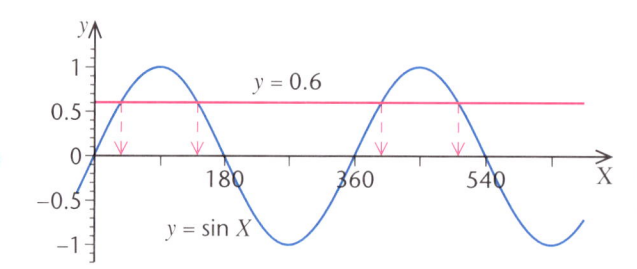

The graph shows that there are four values of $2x - 30°$ in the interval that make the equation work.

Using the symmetry of the curve, the values of $2x - 30°$ are:
$36.86...°, 180° - 36.86...°, 360° + 36.86...°, 540° - 36.86...°$.

This gives:
$2x - 30° = 36.86...° \Rightarrow x = 33.4°$
$2x - 30° = 143.14...° \Rightarrow x = 86.6°$
$2x - 30° = 396.86...° \Rightarrow x = 213.4°$
$2x - 30° = 503.14...° \Rightarrow x = 266.6°$
So $x = 33.4°, 86.6°, 213.4°, 266.6°$ (1 d.p.).

Core 2 Pure Mathematics

Using trigonometric identities

The following trigonometric relationships are **identities**. This means that they are true for all values of x.

They are very important when simplifying expressions and solving equations and should be learnt.

> $$\tan x = \frac{\sin x}{\cos x}$$
>
> **KEY POINT**

Example
Solve the equation $\sin x = \sqrt{3} \cos x$, for values of x between $0°$ and $360°$.

This is allowed since $\cos x = 0$ is not a solution.

Divide each side by $\cos x$.
$$\frac{\sin x}{\cos x} = \sqrt{3}$$
$$\Rightarrow \tan x = \sqrt{3}$$
$$x = 60° \text{ or } 240°$$

The graph of $y = \tan x$ repeats every $180°$, so to find the second solution, add $180°$

> $$\sin^2 x + \cos^2 x = 1$$
>
> **KEY POINT**

This is sometimes known as the Pythagorean trigonometric identity.

Example Solve the equation $\sin^2 x - \cos^2 x - \sin x = 1$ for $-\pi \leqslant x \leqslant \pi$.

Replacing $\cos^2 x$ with $1 - \sin^2 x$ gives $\sin^2 x - (1 - \sin^2 x) - \sin x = 1$
$$\Rightarrow 2\sin^2 x - \sin x - 2 = 0.$$

*You need to recognise that this is a quadratic in $\sin x$.
It has to be rearranged in order to use the quadratic formula.*

Using the formula gives $\sin x = \dfrac{1 \pm \sqrt{1 + 16}}{4} = \dfrac{1 \pm \sqrt{17}}{4}$

$-1 \leqslant \sin x \leqslant 1$ for all values of x.

$\Rightarrow \sin x = 1.28 \ldots$ (no solutions) or $\sin x = -0.78077 \ldots$.

Remember to set your calculator to radians mode.

Using $\sin^{-1}$ gives $\qquad x = -0.8959^c$ to 4 d.p.

Another solution in the interval $-\pi \leqslant x \leqslant \pi$ is $-\pi + 0.8959^c$ giving $x = -2.2457^c$ to 4 d.p.

The diagram shows there are no other solutions in this interval.

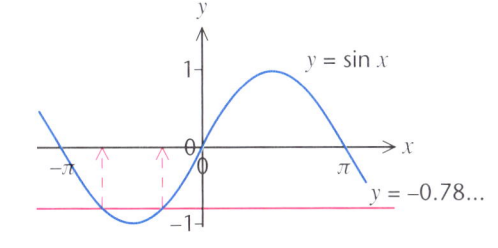

47

Core 2 Pure Mathematics

Progress check

1. Find the exact value of:
   (a) sin 120° (b) cos 135° (c) tan 240°.

2. Describe the transformations needed to turn the graph of $y = \cos x$ into the graph of:
   (a) $y = \cos 2x$ (b) $y = 3 \cos x$ (c) $y = 3 \cos(2x) - 1$.

3. Find the value of x and θ.

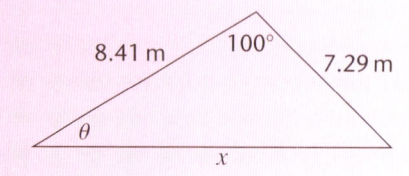

   8.41 m, 100°, 7.29 m, θ, x

4. Find the exact value of:
   (a) $\sin \dfrac{\pi}{4}$ (b) $\cos \dfrac{\pi}{3}$ (c) $\tan \dfrac{\pi}{6}$.

5. Solve $8 \sin x = 3 \cos x$ for $-360° \leq x \leq 360°$.

6. Solve $\cos(3x + 20°) = 0.6$ for $0° \leq x \leq 180°$.

7. Solve $2 \sin^2 x + \cos x = 1$ for $-\pi \leq x \leq \pi$.

8. OAB is a sector of a circle, centre O, radius 4 cm.
   Angle AOB = 0.5 radians.

   (a) Find the perimeter of sector AOB.
   (b) Find the length of the chord AB.
   (c) Find the area of the sector OAB.
   (d) Find the area of the triangle OAB.
   (e) Find the area of the shaded segment.

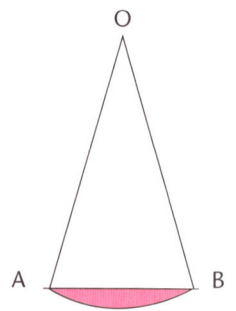

---

1 (a) $\dfrac{\sqrt{3}}{2}$ (b) $-\dfrac{1}{\sqrt{2}}$ (c) $\sqrt{3}$

2 (a) One-way stretch with scale factor 0.5 parallel to the x-axis.
  (b) One-way stretch with scale factor 3 parallel to the y-axis.
  (c) One-way stretch with scale factor 0.5 parallel to the x-axis, followed by a one-way stretch with scale factor 3 parallel to the y-axis, followed by a translation of 1 unit downwards.

3 $x = 12.0$ m (3 s.f.) $\theta = 36.6°$ (1 d.p.).

4 (a) $\dfrac{1}{\sqrt{2}}$ (b) $\dfrac{1}{2}$ (c) $\dfrac{\sqrt{3}}{3}$

5 $\tan x = \dfrac{3}{8}$ giving $x = 200.6°, 20.6°, -159.4°$ and $-339.4°$

6 11.0°, 95.6°, 131.1°

7 $\dfrac{2\pi}{3}, 0, -\dfrac{2\pi}{3}$

8 (a) 10 cm (b) 1.98 cm (2 d.p.) (c) 4 cm² (d) 3.84 cm² (2 d.p.) (e) 0.16 cm² (2 d.p.)

Core 2 Pure Mathematics

2.4 Integration

After studying this section you should be able to:

- find an indefinite integral and understand what it represents
- evaluate a definite integral
- find the area under a curve
- use the trapezium rule to find an approximate value of an integral

LEARNING SUMMARY

Indefinite integration

The idea of a **reverse process** is an important one in many areas of mathematics. In **calculus**, the reverse process of differentiation is **integration** and this turns out to be an extremely important process, in its own right, with many powerful applications.

When you differentiate x^n with respect to x you may think of the process involving two stages:

$$x^n \to \boxed{\text{multiply by the power}} \to \boxed{\text{reduce the power by 1}} \to nx^{n-1}$$

This *suggests* that the reverse process is given by:

$$\frac{x^{n+1}}{n+1} \leftarrow \boxed{\text{divide by the power}} \leftarrow \boxed{\text{increase the power by 1}} \leftarrow x^n$$

However, this doesn't give the *complete* picture. The reason is that you can differentiate $x^n +$ (any constant) and still obtain nx^{n-1}. You need to take this into account when you reverse the process.

$\int$ is the symbol for integration and dx is used to show that the integration is with respect to the variable x.

> **KEY POINT**
> The result is written as $\int x^n \, dx = \frac{x^{n+1}}{n+1} + c$ where c is called the **constant of integration**. Note that $n \neq -1$.

Some examples are:

$$\int x^2 \, dx = \frac{x^3}{3} + c \qquad \int \sqrt{x} \, dx = \int x^{\frac{1}{2}} \, dx = \frac{2}{3} x^{\frac{3}{2}} + c \qquad \int 3 \, dx = 3x + c$$

Notice that $\frac{1}{(\frac{3}{2})} = \frac{2}{3}$.

$$\int \frac{1}{x^2} \, dx = \int x^{-2} \, dx = -x^{-1} + c = -\frac{1}{x} + c.$$

Sums and differences of functions are treated in the same way as in differentiation, by dealing with each term separately. The general rules are given below.

Function notation is useful here but you don't need to state the rules every time you use them.

> **KEY POINT**
> $$\int (f(x) \pm g(x)) \, dx = \int f(x) \, dx \pm \int g(x) \, dx$$
> $$\int a f(x) \, dx = a \int f(x) \, dx \quad \text{(where } a \text{ is a constant)}$$

Example 1

$$\int (3x^2 - 5x + 2) \, dx = x^3 - \tfrac{5}{2} x^2 + 2x + c.$$

49

Core 2 Pure Mathematics

Example 2

$$\int (x^{\frac{3}{2}} + 2x^{-\frac{1}{2}})dx = \frac{x^{\frac{5}{2}}}{\frac{5}{2}} + 2\frac{x^{\frac{1}{2}}}{\frac{1}{2}} + c$$

$$= \frac{2}{5}x^{\frac{5}{2}} + 4x^{\frac{1}{2}} + c$$

Example 3

Write the expression in index form first.

$$\int \frac{x+2}{\sqrt{x}} dx = \int (x^{\frac{1}{2}} + 2x^{-\frac{1}{2}})dx$$

$$= \frac{2}{3}x^{\frac{3}{2}} + 4x^{\frac{1}{2}} + c$$

In some situations you have enough information to find the value of the constant.

Example Find the equation of the curve with gradient function $3x^2$ passing through (2, 5).

Using the point (2, 5) gives $5 = 2^3 + c$ so $c = -3$.

$$\frac{dy}{dx} = 3x^2 \Rightarrow y = \int 3x^2 \, dx \Rightarrow y = x^3 + c. \quad \text{When } x = 2, y = 5 \Rightarrow c = -3.$$

The equation of the curve is $y = x^3 - 3$.

Definite integration

An integral of the form $A = \int_a^b f(x)dx$ is a **definite integral**. It has a numerical value and is evaluated as follows:

Substitute the upper limit, then subtract the value when you substitute the lower limit.

$$\int_a^b f(x)dx = [g(x)]_a^b = g(b) - g(a)$$

KEY POINT

*A constant of integration is **not** used in a definite integral.*

For example $\int_1^4 2x \, dx = [x^2]_1^4 = 4^2 - 1^2 = 15.$

You may need to evaluate a definite integral in which the upper limit is infinite, as in the example below.

$$\int_1^\infty x^{-2} dx = \left[-\frac{1}{x}\right]_1^\infty = \lim_{x \to \infty}\left(-\frac{1}{x}\right) - \left(-\frac{1}{1}\right) = 0 + 1 = 1$$

Area under a curve

The area enclosed by the curve $y = f(x)$, the x-axis and the lines $x = a$ and $x = b$ is given by $\int_a^b y \, dx$.

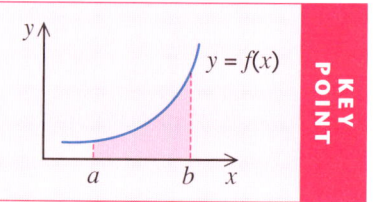

KEY POINT

For areas below the x-axis, the definite integral gives a negative value.

The area between the curve $y = f(x)$, the y-axis and the lines $y = c$ and $y = d$ is given by $\int_c^d x \, dy$.

50

Core 2 Pure Mathematics

Example Find the area under the curve $y = x^2 + 1$ between $x = 1$ and $x = 3$.

$$\text{Area} = \int_1^3 (x^2 + 1)\,dx = \left[\frac{x^3}{3} + x\right]_1^3$$

$$= \left(\frac{3^3}{3} + 3\right) - \left(\frac{1^3}{3} + 1\right)$$

$$= 9 + 3 - \frac{1}{3} - 1$$

$$= 10\tfrac{2}{3}.$$

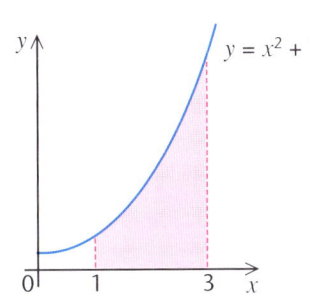

To find the **area between a line and a curve**, find the area under the line and the area under the curve separately and then subtract to find the required area.

Progress check

1. Integrate each of these with respect to x:
   (a) $\dfrac{1}{\sqrt{x}}$
   (b) $\sqrt{x^3}$

2. A curve has gradient function $4x^3 + 1$ and passes through the point $(1, 9)$. Find the equation of the curve.

3. Find the value of each of these definite integrals.
   (a) $\displaystyle\int_0^5 (x-2)\,dx$ (b) $\displaystyle\int_4^9 \sqrt{x}\,dx$ (c) $\displaystyle\int_0^6 y^2\,dy$

4. Find the area under the curve $y = 3\sqrt{x}$ between $x = 1$ and $x = 9$.

1 (a) $2\sqrt{x} + c$ (b) $\tfrac{2}{5}x^{5/2} + c$
2 $y = x^4 + x + 7$
3 (a) 2.5 (b) $12\tfrac{2}{3}$ (c) 72
4 52

Numerical integration – trapezium rule

The value of $A = \displaystyle\int_a^b y\,dx$ represents the area under the graph of $y = f(x)$ between $x = a$ and $x = b$. You can find an approximation to this value using the **trapezium rule**. This is especially useful if the function is difficult to integrate.

The area under the curve between a and b may be divided into n strips of equal width h.

It follows that $h = \dfrac{b-a}{n}$.

Each strip is approximately a trapezium and so the total area is approximately

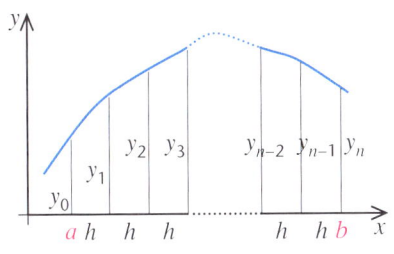

$$\tfrac{1}{2}h(y_0 + y_1) + \tfrac{1}{2}h(y_1 + y_2) + \tfrac{1}{2}h(y_2 + y_3) + \ldots + \tfrac{1}{2}h(y_{n-2} + y_{n-1}) + \tfrac{1}{2}h(y_{n-1} + y_n).$$

To increase the accuracy, use more strips.

> **KEY POINT**
> This simplifies to give the formula known as the **trapezium** rule:
> $$\int_a^b y\,dx = \tfrac{1}{2}h\{(y_0 + y_n) + 2(y_1 + y_2 + \ldots + y_{n-1})\} \text{ where } h = \dfrac{b-a}{n}.$$

Core 2 Pure Mathematics

Example

Use the trapezium rule with five strips to estimate the value of $\int_1^2 2^x \, dx$.

It's useful to tabulate the information:

x	1	1.2	1.4	1.6	1.8	2
2^x	2	2.2974	2.6390	3.0314	3.4822	4
	y_0	y_1	y_2	y_3	y_4	y_5

> Work to a greater level of accuracy in your table than you intend to give in your final answer.

$$h = \frac{2-1}{5} = 0.2$$

So, $\int_1^2 2^x \, dx \approx 0.1\{(2+4) + 2(2.2974 + 2.6390 + 3.0314 + 3.4822)\}$

$= 2.89$ to 3 s.f.

> In this case, the working was done to 5 s.f. and the final answer is given to 3 s.f.

Increasing gradient: over-estimate

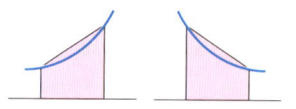

Decreasing gradient: under-estimate

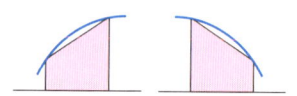

> **KEY POINT**
> If the gradient of the graph is increasing over the interval then the trapezium rule will give an over-estimate of the area. If the gradient of the graph is decreasing over the interval then it will give an under-estimate of the area.

> A graphic calculator with a numerical integration function gives the value of the area as 2.8853901

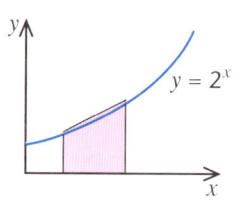

In this case, the gradient is increasing over the interval and so the trapezium rule gives an over-estimate of the area.

Progress check

1 Use the trapezium rule with five strips to estimate $\int_1^2 \frac{1}{x} \, dx$.
   State whether the value obtained is an over-estimate or an under-estimate.

1 0.69563 (5 d.p.)
   Over-estimate

Core 2 Pure Mathematics

Sample questions and model answers

1

Find the exact values of x, where $0 \leqslant x \leqslant 2\pi$, that satisfy the equation
$$\tan x = 2 \sin x.$$

Use the identity $\tan x = \dfrac{\sin x}{\cos x}$

$\tan x = 2 \sin x$

$\dfrac{\sin x}{\cos x} = 2 \sin x$

Multiply both sides by $\cos x$.

$\sin x = 2 \sin x \cos x$

$\sin x - 2 \sin x \cos x = 0$

Do not cancel $\sin x$ as this leads to loss of solutions.

$\sin x (1 - 2 \cos x) = 0$

Either $\sin x = 0 \Rightarrow x = 0, \pi, 2\pi$

or $1 - 2 \cos x = 0 \Rightarrow x = \dfrac{\pi}{3}, \dfrac{5\pi}{3}$

These are exact values.

So $x = 0, \dfrac{\pi}{3}, \pi, \dfrac{5\pi}{3}, 2\pi$

2

(a) Given that $\log_a x - \log_a 8 + 2 \log_a 4 = 0$, where a is a positive constant, find x.

(b) Find the value of x, correct to two decimal places, that satisfies the equation $3^{2x+1} = 40$.

(a) $\log_a x - \log_a 8 + 2 \log_a 4 = 0$

Simplify, using the log laws.

$\Rightarrow \log_a x - \log_a 16 + \log_a 8 = 0$

$\Rightarrow \log_a \left(\dfrac{16x}{8} \right) = 0$

$\Rightarrow \log_a (2x) = 0$

Remember that $a^0 = 1$ for all values of a.

$\Rightarrow 2x = 1$

$x = 0.5$

(b) $3^{2x+1} = 40$

Take logs to the base 10 of both sides.

$\log_{10}(3^{2x+1}) = \log_{10} 40$

$(2x + 1) \log_{10} 3 = \log_{10} 40$

Use $\log_a(b^c) = c \log_a b$

$2x + 1 = \dfrac{\log_{10} 40}{\log_{10} 3} = 3.3577\ldots$

$2x = 2.3577\ldots$

$x = 1.18 \text{ (2 d.p.)}$

Core 2 Pure Mathematics

Sample questions and model answers (continued)

3

The third term of a geometric series is 20 and the fifth term is 5.
Given that the common ratio, r, is negative, find
(a) the value of r
(b) the first term
(c) the sum to infinity of the series.

The nth term of a geometric series is ar^{n-1}.

(a) $ar^2 = 20$ (1)
$ar^4 = 5$ (2)
Dividing (2) by (1) gives $r^2 = \frac{1}{4}$
$r < 0 \Rightarrow r = -\frac{1}{2}$.

(b) Substitute into (1) $a(\frac{1}{4}) = 20$
$a = 80$

The series is convergent, since $|r| < 1$.

(c) $S_\infty = \dfrac{a}{1-r}$
$= \dfrac{80}{1-(-\frac{1}{2})}$
$= 53\frac{1}{3}$

4

(a) Express $\dfrac{5x^2 - 1}{\sqrt{x}}$ in the form $5x^a - x^b$, where a and b are rational numbers to be found.

(b) Hence find the exact value of $\displaystyle\int_1^2 \dfrac{5x^2 - 1}{\sqrt{x}}\, dx$.

Remember that $\sqrt{x} = x^{1/2}$ and work in index form.

(a) $\dfrac{5x^2 - 1}{\sqrt{x}} = \dfrac{5x^2}{x^{1/2}} - \dfrac{1}{x^{1/2}} = 5x^{3/2} - x^{-1/2}$

So $a = \frac{3}{2}$ and $b = -\frac{1}{2}$.

(b) $\displaystyle\int_1^2 \dfrac{5x^2 - 1}{\sqrt{x}}\, dx = \int_1^2 (5x^{3/2} - x^{-1/2})\, dx$

$= \left[\dfrac{5x^{5/2}}{\frac{5}{2}} - \dfrac{x^{1/2}}{\frac{1}{2}} \right]_1^2$

$= [2x^{5/2} - 2x^{1/2}]_1^2$

$(2)^{5/2} = (\sqrt{2})^5 = 4\sqrt{2}$

$= 8\sqrt{2} - 2\sqrt{2} - (2 - 2)$

Leave the answer in surd form.

$= 6\sqrt{2}$

Practice examination questions

1. The 10th term of an arithmetic progression is 74 and the sum of the first 20 terms is 1510.
   Find the first term and the common difference.

2. (a) Find the 8th term of a geometric progression with first term 3 and common ratio 2.

   (b) Find the sum of the first 50 terms of the series $25 + 26.2 + 27.4 + ...$

3. (a) Given that $(x + 2)$ is a factor of $f(x) = x^3 - 4x^2 - 3x + k$, use the factor theorem to find the value of k.

   (b) Factorise $f(x)$ completely and sketch the graph of $y = f(x)$.

   (c) Solve the inequality $f(x) \geq 0$.

   (d) Find the remainder when $f(x)$ is divided by $(x + 1)$.

4. (a) Express $3 \log_a x - 2 \log_a y + \log_a(x + 1)$ as a single logarithm.

   (b) Solve the equation $5^x = 100$. Give your answer to 3 d.p.

Practice examination questions (continued)

5. Solve the equation $\cos(2x + 30°) = 0.4$ for $0° \leq x \leq 360°$.

6. (a) Write the equation $\cos x + 3 \sin x \tan x - 2 = 0$ in the form $a \cos^2 x + b \cos x + c = 0$.

   (b) Find the solutions of $\cos x + 3 \sin x \tan x - 2 = 0$ for $0° \leq x \leq 360°$.

7. A function $f(x)$ is defined by:

   $f(x) = x^3 + 4x^2 - 3x - 18$

   for all real values of x.

   (a) Given that $(x - 2)$ is a factor of $f(x)$, factorise $f(x)$ completely.

   (b) Sketch the graph of $y = f(x)$.

   (c) Sketch the graphs of $y = f(x) + 10$ and $y = f(x - 2)$.

Practice examination questions (continued)

8 The diagram shows a region R bounded by the curve $y = 3\sqrt{x}$, the x-axis and the lines $x = 1$ and $x = 4$.

Calculate the area of the region R.

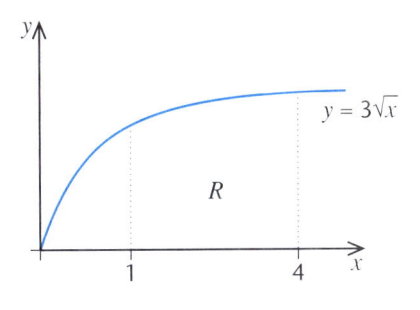

9 The diagram shows a circle of unit radius. The angle between the radii is θ where $0 < \theta < \pi$ and the area of the shaded segment is $\dfrac{\pi}{5}$.

Show that $\theta = \sin \theta + \dfrac{2\pi}{5}$.

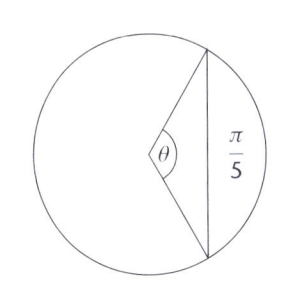

10 The area shaded in the diagram is bounded by the curve $y = \sqrt{x^2 - 3}$, the lines $x = 2$ and $x = 3$ and the x-axis.

Express the shaded area as an integral and estimate its value using the trapezium rule with five intervals. Give your answer to 2 d.p.

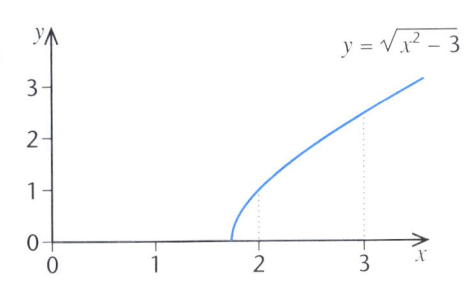

Chapter 3
Mechanics 1

The following topics are covered in this chapter:

- Kinematics
- Statics
- Dynamics

3.1 Kinematics

LEARNING SUMMARY

After studying this section you should be able to:
- understand the distinction between vector and scalar quantities
- apply the constant acceleration formulae for motion in a straight line
- draw and interpret graphs of displacement, velocity and acceleration against time
- use calculus for motion with variable acceleration

Vector and scalar quantities

A **scalar** quantity has size (or **magnitude**) but not direction. **Numbers** are scalars and some other important examples are **distance**, **speed**, **mass** and **time**.

A **vector** quantity has both size and **direction**. For example, distance in a specified direction is called **displacement**. Some other important examples are **velocity**, **acceleration**, **force** and **momentum**.

The diagram shows a **directed line segment**. It has size (in this case, length) and direction so it is a vector.

The diagram gives a useful way to represent *any* vector quantity and may also be used to represent addition and subtraction of vectors and multiplication of a vector by a scalar.

In a textbook, vectors are usually labelled with lower case letters in **bold** print.
When hand-written, these letters should be underlined e.g. $\underline{a}$

Motion in a straight line

You can use these formulae whenever the acceleration of an object is constant.

$$v = u + at$$
$$s = ut + \tfrac{1}{2}at^2$$
$$s = \left(\frac{u+v}{2}\right)t$$
$$v^2 = u^2 + 2as$$

s is the displacement from a fixed position
u is the initial velocity
a is the acceleration
t is the time that the object has been in motion
v is the velocity at time t.

Example
An object starts from rest and moves in a straight line with constant acceleration 3 m s^{-2}. Find its velocity after 5 s.

From the given information:
$u = 0$
$a = 3$
$t = 5$.

Mechanics 1

Don't include units in your working.

Make a clear statement and include the appropriate units.

Using $v = u + at$

gives $v = 0 + 3 \times 5 = 15$.

The velocity after 5 s is 15 m s^{-1}.

> **KEY POINT**
> Motion may take place in either direction along a straight line. One direction is taken to be positive for displacement, velocity and acceleration and the other direction is taken to be negative.

In some questions the acceleration due to gravity is taken to be 9.8 m s^{-2}.

Example

A stone is thrown vertically upwards with a velocity of 20 m s^{-1}. It has a downward acceleration due to gravity of 10 m s^{-2}.

(a) Find the greatest height reached by the stone.
(b) Find its velocity after 3 s.
(c) Find the height of the stone above the point of projection after 3 s.

(a) From the given information: $u = 20$
$a = -10$.

At the greatest height $v = 0$.
Using $v = u + at$
$0 = 20 - 10t \Rightarrow t = 2$.

Using $s = ut + \frac{1}{2}at^2$
gives $s = 20 \times 2 - 5 \times 4 = 20$

The greatest height reached is 20 m.

(b) Using
$v = u + at$ with $t = 3$
gives
$v = 20 - 10 \times 3 = -10$.

After 3 s the stone is moving downwards at 10 m s^{-1}.

(c) Using
$s = ut + \frac{1}{2}at^2$
gives
$s = 20 \times 3 - 5 \times 3^2 = 15$.

After 3 s the stone is 15 m above the point of projection.

Graphical representation

You need to know the properties of the graphs of distance, displacement, speed, velocity and acceleration against time.

> **KEY POINT**
> The gradient of a distance–time graph represents speed.
> The gradient of a displacement–time graph represents velocity.
> The gradient of a velocity–time graph represents acceleration.
>
> The area under a speed–time graph represents the distance travelled.
> The area under a velocity–time graph represents the change of displacement.
> The area under an acceleration–time graph represents change in velocity.

Mechanics 1

Example
The diagram represents the progress of a car as it travels between two sets of traffic lights.
Find:

(a) The initial acceleration of the car.

(b) The deceleration of the car as it approaches the second set of lights.

(c) The distance between the traffic lights.

(a) Gradient of OA = $\frac{15}{10}$ = 1.5

The initial acceleration of the car is 1.5 m s^{-2}.

(b) Gradient of BC = $-\frac{15}{15}$ = -1.

The *acceleration* of the car is -1 m s^{-2} so the *deceleration* is 1 m s^{-2}.

(c) The area of the trapezium OABC is given by $\frac{15}{2}(5 + 30)$ = 262.5, so, the distance between the traffic lights is 262.5 m.

Variable acceleration

In this case, the constant acceleration formulae do not apply and must not be used. Using x for displacement, v for velocity and a for acceleration:

$$\text{differentiate} \xrightarrow{x \quad v \quad a} \qquad v = \frac{dx}{dt} \qquad a = \frac{dv}{dt} = \frac{d^2x}{dt^2}$$

$$\text{integrate} \xleftarrow{} \qquad x = \int v\,dt \qquad v = \int a\,dt.$$

Example

A particle P moves along the x-axis such that its velocity at time t is given by $v = 5t - t^2$. When $t = 0$, $x = 15$. Find a formula for:

(a) the acceleration of the particle at time t.

(b) the position of the particle at time t.

(a) $v = 5t - t^2 \Rightarrow \frac{dv}{dt} = 5 - 2t$.

The acceleration of the particle at time t is given by $a = 5 - 2t$.

(b) $\qquad x = \int v\,dt = \int 5t - t^2\,dt$

$\Rightarrow \qquad x = \frac{5t^2}{2} - \frac{t^3}{3} + c.$

When $t = 0$, $x = 15$

so $15 = 0 + c \Rightarrow c = 15$.

This gives the position of the particle at time t as $x = \frac{5t^2}{2} - \frac{t^3}{3} + 15$.

Mechanics 1

Progress check

1 A stone is thrown vertically upwards with a speed of 15 m s^{-1}. Take the downward acceleration due to gravity to be 10 m s^{-2} and find:
   (a) The greatest height reached.
   (b) The speed and direction of the stone after 2 s.

2 A particle P moves along the x-axis such that its displacement x metres from its starting point at time t seconds is given by:

$x = t(t - 1)$

   (a) Find the velocity of the particle when $t = 5$ s.
   (b) Find the speed of the particle and its direction of movement when it returns to its initial position.

1 (a) 11.25 m
  (b) 5 m s^{-1} downwards
2 (a) $v = 9$ m s^{-1}
  (b) $x = 0$ when $t = 0$ and when $t = 1$, so the particle returns to the starting point when $t = 1$.
  $v = 2t - 1$
  When $t = 1$, $v = 1$.
  The speed of the particle is 1 m s^{-1} in the positive direction of the x-axis.

Mechanics 1

3.2 Statics

After studying this section you should be able to:

- resolve a single force into perpendicular components
- resolve a system of forces in a given direction
- find the resultant of a system of forces
- solve problems involving friction
- apply the conditions for equilibrium of coplanar forces in simple cases

Force

Force is a vector quantity that influences the motion of an object. It is measured in newtons (N). For example, the **weight** of an object is the force exerted on it by gravity. An object with a mass of m kg has weight mg N. **Tension**, **reaction** and **friction** are other examples of force and will be met in this section.

Resolving forces

Here are some examples of resolving forces into two components at right-angles to each other.

In each case the forces are represented in magnitude and direction by the sides of the triangle. The force to be resolved is *always* shown on the hypotenuse.

Notice the directions of the forces shown by the arrows. The diagram represents vector addition of the components.

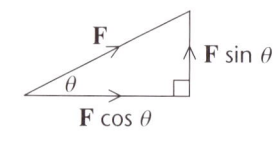

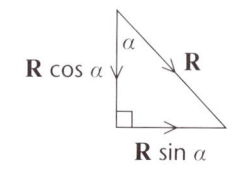

> **KEY POINT**
> These are **vector diagrams** showing the relationship between a force and its components.

In a **force diagram**, you can *replace* a force with a pair of components that are equivalent to it. This will often make it easier to produce the equations necessary to solve a problem.

*Take care not to show both the force and its components on a **force diagram** or you will represent the force TWICE.*

An important example is that of an object on an **inclined plane**.

The diagram shows an object of weight W on a smooth plane inclined at angle θ to the horizontal. R is the force that the plane exerts on the object. It acts at right-angles to the plane and is called the **normal reaction**.

To analyse the behaviour of the object the weight is usually resolved into components parallel and perpendicular to the plane.

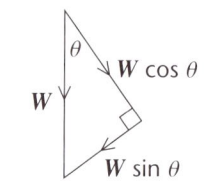

It's worth remembering these results.

This vector diagram shows the component of weight acting down the plane is $W \sin \theta$ and the component perpendicular to the plane is $W \cos \theta$.

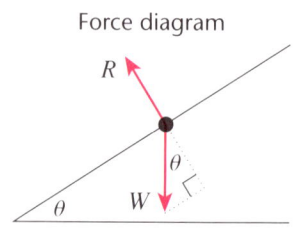

Force diagram

Vector diagram

Mechanics 1

Example
An object of weight 10 N is held in place on a plane inclined at 30° to the horizontal by two forces F N and R N as shown. Find the values of F and R.

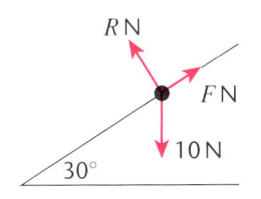

The forces F and R correspond to the components of the weight, parallel and perpendicular to the plane, but act in the opposite directions.

The forces must balance in each direction.

Resolving parallel to the plane: $F = 10 \sin 30° = 5$.

Resolving perpendicular to the plane: $R = 10 \cos 30° = 5\sqrt{3}$.

Resultant force

The effect of several forces acting at a point is the same as the effect of a single force called the **resultant force**. You can find the resultant of a set of forces by using vector addition.

Example
The diagram shows two forces acting at a point on an object. Find the magnitude of the resultant force.

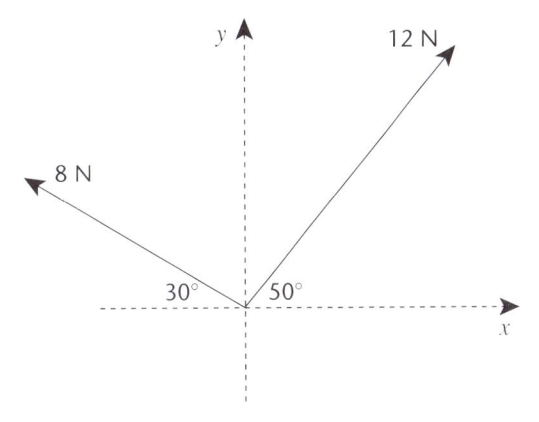

Resolving horizontally $12\cos 50° - 8\cos 30° = 0.78524...$

Resolving vertically $8\sin 30° + 12\sin 50° = 13.192...$

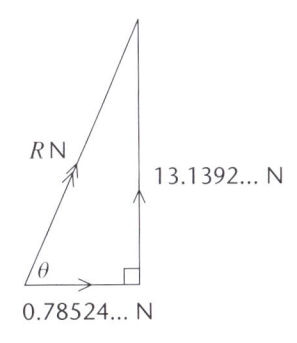

$R = \sqrt{(0.78524^2 + 13.1392^2)} = 13.2$ (3 s.f.)

$\tan \theta = \dfrac{13.1392}{0.78524}$

$\theta = 86.6°$ (3 s.f.)

The resultant force is 13.2 N in the direction 86.6° anticlockwise from the positive x-axis.

Mechanics 1

Equilibrium

A set of forces acting at a point is in **equilibrium** if the resultant force is zero.

It follows that the resolved parts of the forces must balance in any chosen direction.

Example

The forces given in the previous example are shown again here. Find the magnitude and direction of a third force, F, such that the three forces are in equilibrium.

For equilibrium, F must have the same magnitude as the resultant of the two forces, but lie in the opposite direction.

F has magnitude 13.2 N in the direction 93.4° clockwise from the positive x-axis.

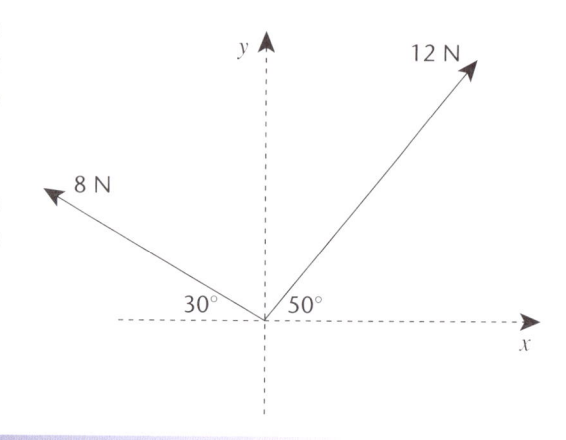

Friction

Whenever two rough surfaces are in contact, the tendency of either surface to move relative to the other is opposed by the **force of friction**.

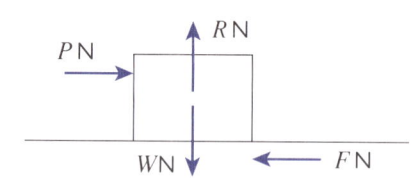

The diagram shows an object of weight W N resting on a rough horizontal surface. The object is pushed from one side by a force P N and friction responds with force F N in the opposite direction.

For small values of P, no movement takes place and $F = P$.

If P increases then F increases to maintain equilibrium until F reaches a maximum value. At this point the object is in **limiting equilibrium** and is on the point of slipping.

If P is now increased again then F will remain the same. The equilibrium will be broken and the object will move.

The maximum value of F depends on:

- The magnitude of the normal reaction R.
- The roughness of the two surfaces measured by the value μ. This value is known as the **coefficient of friction**.

In general: $\qquad\qquad F \leqslant \mu R$.
For limiting equilibrium: $\qquad F = \mu R$.

Mechanics 1

Example
An object of mass 8 kg rests on a rough horizontal surface. The coefficient of friction is 0.3 and a horizontal force P N acts on the object which is about to slide. Take $g = 10$ m s^{-2} and find the value of P.

Resolving vertically gives $R = 80$.

Friction is limiting so $F = \mu R = 0.3 \times 80 = 24$.

Resolving horizontally: $P = F \Rightarrow P = 24$.

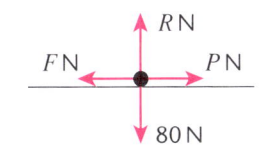

> Remember that the weight of the object is given by mg.

Example
In the diagram, the object of mass m kg is on the point of slipping down the plane. The coefficient of friction between the object and the plane is μ. Show that $\mu = \tan \theta$.

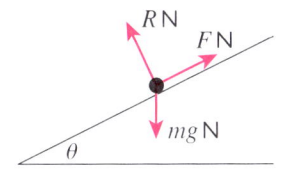

Resolving perpendicular to the plane $\qquad R = mg \cos \theta$.
Resolving parallel to the plane $\qquad F = mg \sin \theta$.
Friction is limiting so $\qquad F = \mu R$.

This gives $\qquad mg \sin \theta = \mu mg \cos \theta \Rightarrow \mu = \dfrac{\sin \theta}{\cos \theta} = \tan \theta$.

Progress check

1. An object of mass 8 kg rests on a plane inclined at 40° to the horizontal. Find the components of its weight parallel and perpendicular to the plane. Take $g = 9.81$ m s^{-2}.

2. Find the magnitude and direction of the resultant of the two forces shown.

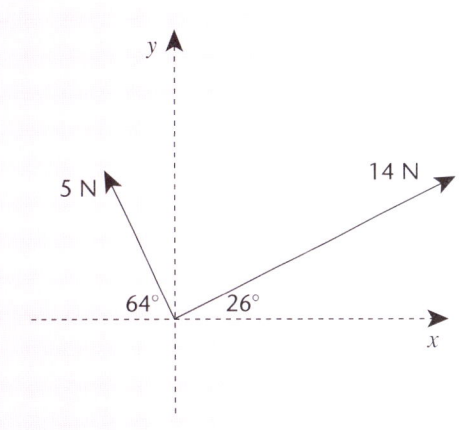

3. An object of weight 50 N rests on a rough horizontal surface. A horizontal force of 20 N is applied to the object so that it is on the point of slipping. Find the value of the coefficient of friction.

1 parallel: 50.4 N, perp: 60.1 N
2 14.9 N at 45.7° anticlockwise to the positive x-axis.
3 0.4

Mechanics 1

3.3 Dynamics

After studying this section you should be able to:

- understand and apply Newton's laws of motion in two or three dimensions
- analyse the motion of connected particles
- understand and apply the principle of conservation of momentum

Newton's laws of motion

Newton's laws of motion provide us with a clear set of rules that can be used to analyse the effect of forces within a system.

The laws may be stated as:
1. Every particle continues in a state of rest or uniform motion, in a straight line, unless acted upon by an external force.
2. The resultant force acting on a particle is equal to its rate of change of momentum. *This law is most often applied in the form $F = ma$.*
3. Every force has an equal and opposite reaction.

The formula F = ma

In the formula $F = ma$, F N stands for the resultant force acting on a particle, m kg is the mass of the particle and a m s^{-2} is the acceleration produced.

It is important to use the correct units.

Example
A particle of mass 2 kg rests on a smooth horizontal plane.
Horizontal forces of 15 N and 4 N act on the particle in opposite directions.
Find the acceleration of the particle.

Using $F = ma$

$$15 - 4 = 2a$$
$$\Rightarrow a = 5.5$$

The acceleration of the particle is 5.5 m s^{-2}.

Connected particles

In a typical problem, two particles are connected by a **light inextensible string**.
Since the string is light, there is no need to consider its mass.
Since it is inextensible, both particles will have the same speed and accelerate at the same rate *while the string is taut*.

By Newton's third law, the tension in the string acts equally on both particles but in opposite directions.

To solve problems involving connected particles you need to:

- Draw a diagram.
- Identify and label the forces acting on each particle.
- Write the **equation of motion** for each particle, i.e. apply $F = ma$ for each one.
- Solve the simultaneous equations produced.

Mechanics 1

Example
Two particles P and Q are connected by a light inextensible string. The particles are at rest on a smooth horizontal surface and the string is taut. A force of 10 N is applied to particle Q in the direction PQ. P has mass 2 kg and Q has mass 3 kg. Find the tension in the string and the acceleration of the system.

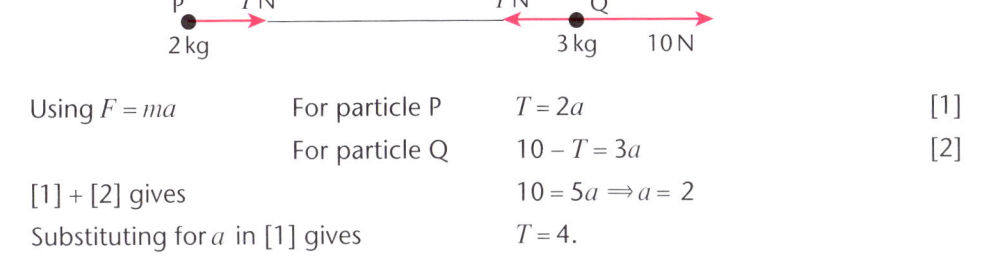

Note the use of the double headed arrow to represent acceleration.

[Diagram: P (2 kg) with T N arrow right; Q (3 kg) with T N arrow left and 10 N arrow right; acceleration a m s^{-2} to the right]

Using $F = ma$ For particle P $T = 2a$ [1]

For particle Q $10 - T = 3a$ [2]

[1] + [2] gives $10 = 5a \Rightarrow a = 2$

Substituting for a in [1] gives $T = 4$.

The tension in the string is 4 N and the acceleration of the system is 2 m s^{-2}.

Momentum

The **momentum** of a particle is a vector quantity given by the product of its mass and its velocity, i.e. momentum = $m\mathbf{v}$ where m kg is the mass of a particle and $\mathbf{v}$ m s^{-1} is its velocity.

The principle of conservation of momentum states that when two particles collide:

the total momentum before impact = the total momentum after impact.

This kind of diagram gives a useful way to represent the information.

Before impact: A (m_1 kg) with velocity u_1 m s^{-1}; B (m_2 kg) with velocity u_2 m s^{-1}

After impact: v_1 m s^{-1} and v_2 m s^{-1}

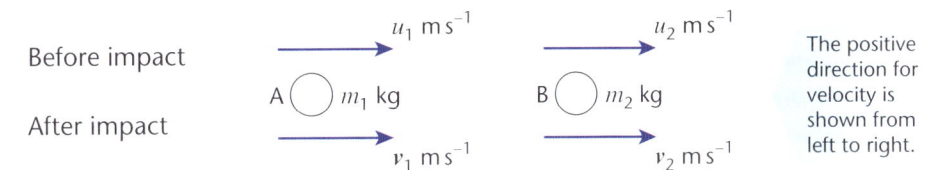

The positive direction for velocity is shown from left to right.

In the diagram, $u_1 > u_2$ so that the particles collide.

Conservation of momentum gives $m_1 u_1 + m_2 u_2 = m_1 v_1 + m_2 v_2$.

Example
A particle of mass 5 kg moving with speed 4 m s^{-1} hits a particle of mass 2 kg moving in the opposite direction with speed 3 m s^{-1}. After the impact the two particles move together with the same speed v m s^{-1}. Find the value of v.

Unknown velocities are shown in the positive direction.

Before impact: 5 kg at 4 m s^{-1} (right); 2 kg at 3 m s^{-1} (left)

After impact: v m s^{-1} and v m s^{-1}

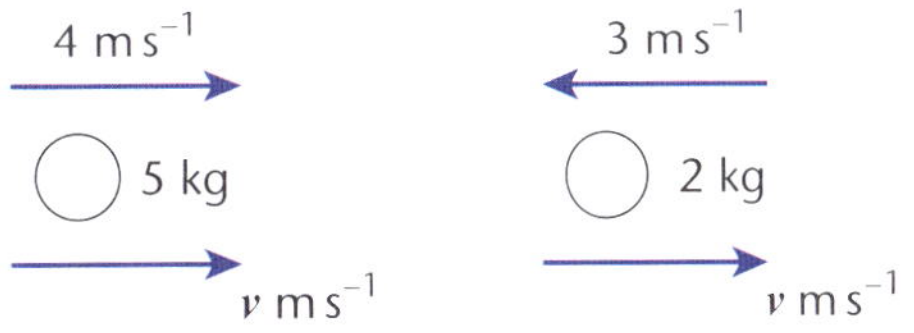

Conservation of momentum gives $5 \times 4 - 2 \times 3 = 5v + 2v$

$\Rightarrow \quad 7v = 14$

$\Rightarrow \quad v = 2$.

Mechanics 1

Progress check

1. A particle of mass 2 kg is acted upon by a force of 5 N in the direction of the positive *x*-axis and a force of 12 N in the direction of the positive *y*-axis.
   (a) Find the magnitude and direction of the resultant force.
   (b) Find the magnitude and direction of the acceleration.

2. A particle moves along the *x*-axis such that its displacement *x* m from O at time *t* s is given by $x = t^3 - 15$ for $0 \leqslant t \leqslant 5$.
   (a) Find the acceleration of the particle at time *t* s.
   (b) Given that the particle has mass 5 kg find the resultant force acting on it when $t = 4$.

3. Two particles A and B are connected by a light inextensible string. The particles are at rest on a smooth horizontal surface and the string is taut. A force of 12 N is applied to particle B in the direction AB. A has mass 4 kg and B has mass 2 kg. Find the tension in the string and the acceleration of the system.

1. (a) Resultant force is 13 N in direction 67.4° anti-clockwise from positive *x* direction.
   (b) Acceleration is 6.5 m s^{-2} in same direction as resultant force.
2. (a) 6*t* m s^{-2}
   (b) 120 N.
3. Tension = 8 N, acceleration = 2 m s^{-2}.

Mechanics 1

Sample questions and model answers

1

A particle moving in a straight line passes through a fixed point O when $t = 0$. Its velocity at time t seconds is given by $v = 12 - 3t^2$ m s^{-1}.

(a) Find the acceleration of the particle when $t = 4$.

(b) Find the distance of the particle from O:

  (i) when it comes to rest
  (ii) when $t = 5$.

(a) $v = 12 - 3t^2 \implies a = \dfrac{dv}{dt} = -6t$

$t = 4 \implies a = -6 \times 4 = -24.$

The acceleration of the particle when $t = 4$ is -24 m s^{-2}.

(b) (i) The particle comes to rest when $v = 0$.

$v = 0 \implies 12 - 3t^2 = 0$

$\implies t = 2$ since $t > 0$.

The displacement of the particle from O at time t is given by

$$x = \int v\, dt + c = \int 12 - 3t^2\, dt + c$$

$\implies x = 12t - t^3 + c.$

When $t = 0$, $x = 0$

giving $\qquad 0 = 0 + c \implies c = 0.$

So $\qquad x = 12t - t^3.$

When $t = 2 \qquad x = 12 \times 2 - 2^3 = 16.$

The particle is 16 m from O when it comes to rest.

(b) (ii) When $t = 5 \qquad x = 12 \times 5 - 5^3 = -65.$

The particle is 65 m from O when $t = 5$.

Distance is taken to be positive.

Mechanics 1

Sample questions and model answers (continued)

2

Two particles are connected by a light inextensible string passing over a light frictionless pulley. Particle A has mass 5 kg and lies on a smooth horizontal table. Particle B has mass 4.5 kg and hangs freely. Take $g = 10$ m s^{-2}.

(a) Write down the equation of motion for each particle.

(b) Find the acceleration of the system.

(c) Find the tension in the string.

(a)

A clearly labelled diagram is an essential first step.

Both particles have the same acceleration.

The tension is the same at both ends of the string.

[Diagram: 5 kg block A on table with tension T N and acceleration a m s^{-2} to the right; string passes over pulley to 4.5 kg block B hanging, with tension T N up and weight 45 N down.]

The diagram shows the forces acting on each particle in the direction of motion.

The equations of motion are:

Use F = ma for each particle in turn.

for particle A $T = 5a$ [1]

for particle B $45 - T = 4.5a$ [2]

Solve the simultaneous equations and interpret the results.

(b) [1] + [2] gives $45 = 9.5a$

$\Rightarrow a = 4.74$ to 3 s.f.

The acceleration of the system is 4.74 ms^{-2} to 3 s.f.

(c) Substitution for a in [1] gives $T = 23.7$ to 3 s.f.

It is a good idea to state the degree of accuracy given in your answer.

The tension in the string is 23.7 N to 3 s.f.

Mechanics 1

Practice examination questions

1 A particle moving in a straight line passes through a point O with velocity 10 m s^{-1} when $t = 0$. Given that the acceleration of the particle is -4 m s^{-2}, find:

(a) The velocity of the particle when $t = 5$.

(b) The distance that the particle travels between $t = 0$ and $t = 4$.

2

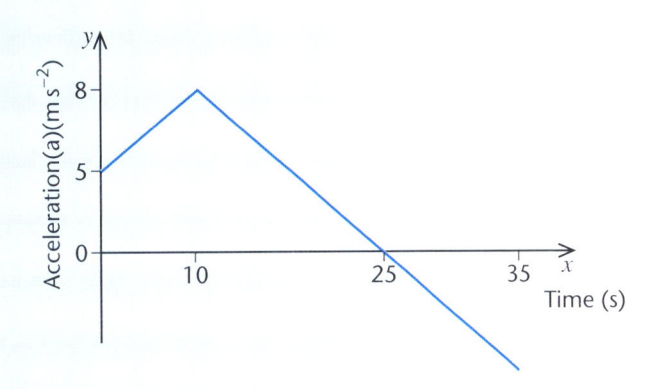

The diagram is an acceleration–time graph for the motion of a particle during a period of 35 s. The particle moves in a straight line and its initial velocity is 20 m s^{-1}.

(a) State the time at which the velocity reaches its maximum value.

(b) Calculate the maximum value of the velocity of the particle during this period.

Mechanics 1

Practice examination questions (continued)

3

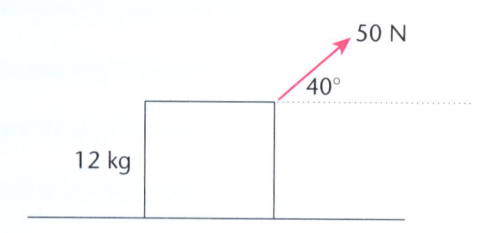

The diagram shows an object of mass 12 kg resting on a rough horizontal surface.

A force of 50 N acts on the object at an angle of 40° to the horizontal.
Take $g = 10$ m s^{-2}.

(a) Find the normal reaction between the horizontal surface and the object.

(b) Given that the object is in limiting equilibrium, find the coefficient of friction between the object and the horizontal surface.

4 The diagram shows a mass of 6 kg acted upon by a horizontal force of P N.

The coefficient of friction between the object and the plane is 0.3

Find the minimum value of P required to maintain the object in equilibrium.

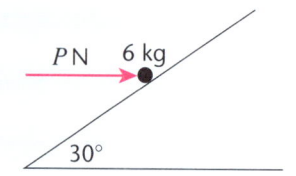

Practice examination questions (continued)

5

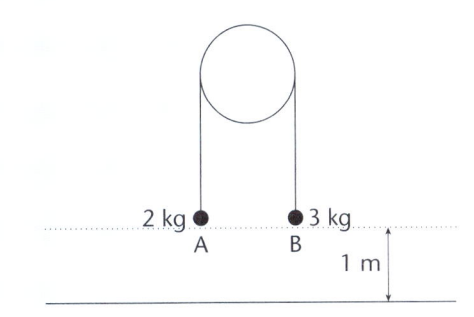

2 kg • A   • 3 kg B   1 m

Particles A and B of mass 2 kg and 3 kg respectively are connected by a light inextensible string passing over a smooth frictionless pulley. The system is released from rest with both particles 1 m above the ground. Find:

(a) The acceleration of the system while the string remains taut.

(b) The tension in the string.

(c) The time taken for particle B to reach the ground.

(d) The speed with which particle B hits the ground.

(e) The greatest height reached by particle A.

(f) What difference would it make to your initial equations if the pulley was not light and frictionless?

Mechanics 1

Practice examination questions (continued)

6 A particle moves along the x-axis such that at time t seconds its displacement from a point O on the line is $x = t^3 - 6t^2 + 3$ metres.

   (a) Show that the particle is initially at rest and find its greatest distance from O in the negative direction.

   (b) Find the speed of P when it passes through its initial position.

7 A particle P moves along a straight line passing through the point O. The displacement x metres, of the particle from O at time t seconds is given by:

   $x = 4t^3 - 3t^2 + 7$ for $0 \leqslant t \leqslant 2$.

   Find:

   (a) the velocity of the particle at time t

   (b) the distance that the particle travels before it changes direction

   (c) the time taken for the particle to return to its starting point.

Chapter 4
Probability and Statistics 1

The following topics are covered in this chapter:

- Representing data
- Probability
- Discrete random variables
- Correlation and regression

4.1 Representing data

After studying this section you should be able to:

- calculate averages for discrete and continuous data
- find the variance, standard deviation, range and interquartile range of data
- draw diagrams to represent and compare distributions
- interpret measures of location and dispersion in comparing sets of data
- understand the concepts of outliers and skewness

Measures of central location

An average is a value that is taken to be representative of a data set. The three forms of average that you need are the mean, median and mode. These are often referred to as **measures of central location**.

The **mean** $\bar{x}$ of the values $x_1, x_2, x_3, ..., x_n$ is given by $\bar{x} = \dfrac{\sum x_i}{n}$. If each x_i occurs with frequency f_i then the mean of the **frequency distribution** is given by $\bar{x} = \dfrac{\sum f_i x_i}{\sum f_i}$.

Example
Find the mean of these results obtained by throwing a dice.

score	1	2	3	4	5	6
frequency	18	17	23	20	24	18

One problem with the mean is that it may be unduly influenced by a small number of extreme values, known as **outliers**.

$$\bar{x} = \dfrac{18 \times 1 + 17 \times 2 + 23 \times 3 + 20 \times 4 + 24 \times 5 + 18 \times 6}{18 + 17 + 23 + 20 + 24 + 18} = \dfrac{429}{120} = 3.575$$

The **median** is the middle value of the data when it is arranged in order of size. If there is an even number of values then the median is the mean of the middle pair. In the example above, the 60th and 61st values are both 4 so the median is 4.

The **mode** is the value that occurs with the highest frequency. In the example above the mode is 5.

You can *estimate* the mean of **grouped data** by using the midpoint of each class interval to represent the class.

Example
Estimate the mean value of h from the figures given in the table.
An estimate of the mean is given by

$$\bar{x} = \dfrac{755}{72} = 10.486...$$

$10.5 =$ to 1 d.p.

Once data have been grouped, the exact values are not available and so it is only possible to *estimate* the mean.

Interval	frequency (f_i)	midpoint (x_i)	$f_i \times x_i$
$0 < h \leqslant 5$	8	2.5	20
$5 < h \leqslant 10$	24	7.5	180
$10 < h \leqslant 15$	29	12.5	362.5
$15 < h \leqslant 20$	11	17.5	192.5
Totals	72		755

Probability and Statistics 1

Measures of dispersion

An average alone gives no indication of how widely dispersed the data values are. The simplest measure of dispersion, or spread, is the **range**.

The range of a data set is the difference between largest value and the smallest value.

A slightly more refined approach is to measure the spread of the 'middle half' of the data. The data is put into order and the values Q1, Q2 and Q3 are found that divide the data into quarters.

> The diagram illustrates the situation for a data set of 11 values.

• • • • • • • • • • •
     Q1   Q2   Q3

Q2 is the **median** of the data.

Q1, the **lower quartile**, is the median of the lower half of the data.

Q3, the **upper quartile**, is the median of the upper half of the data.

The **interquartile range** is then Q3 − Q1.

The quartiles may also be used to indicate whether the data values show **positive skew** (Q2 − Q1 < Q3 − Q2) or **negative skew** (Q2 − Q1 > Q3 − Q2).

The **variance** of a data set is the mean of the squares of the deviations from the mean.

> The second form is easier to work with from raw data.

$$\text{Variance} = \frac{\sum (x_i - \bar{x})^2}{n} = \frac{\sum x_i^2}{n} - \bar{x}^2.$$

The variance gives an indication of the spread of data values about the mean but the units of the data have been squared in the process.

> The value may be obtained directly from a scientific calculator but you may need to apply the formula using summary data. See page 86.

$$\text{Standard deviation} = \sqrt{\text{variance}} = \sqrt{\frac{\sum x_i^2}{n} - \bar{x}^2}.$$

Standard deviation is measured in the same units as the data and is the value most commonly used to measure dispersion at this level.

For grouped data, take the mid-interval value for x.

Statistical diagrams

The use of statistical diagrams may help in comparing distributions and revealing further information about the data.

A **box and whisker plot** for example shows the location and spread of a distribution at a glance.

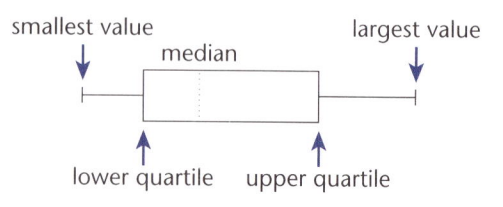

(smallest value, median, largest value, lower quartile, upper quartile)

> Stem and leaf diagrams have the advantage over bar charts that details of the data are available.

A **back-to-back stem and leaf diagram** allows direct comparison of two sets of data to be made.

The diagram gives a sense of location and spread for each data set.

The spread of marks is similar for the boys and the girls but the average for the girls is higher than for the boys.

boys		coursework marks		girls
	2	4	1	7
	5 3	3	2	5 8 8
5 4 4 2 1		2	4	4 7 9
8 7 6 4		1	3	6 6
6 5		0	7	

2 | 4 means 42%   4 | 1 means 41%

Probability and Statistics 1

Histograms are useful for illustrating grouped continuous data.
The area of a bar is proportional to the frequency in that interval.
The height of the rectangle is given by the **frequency density**, where

$$\text{frequency density} = \frac{\text{frequency}}{\text{width of interval}}$$

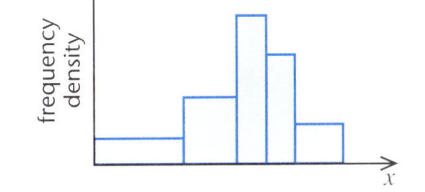

The **modal class** is the interval with the greatest frequency density. This is the highest bar on the histogram.

Joining with straight lines the midpoints of the tops of the bars of a histogram gives a **frequency polygon**. Comparisons can be made by superimposing more than one frequency polygon on the same diagram.

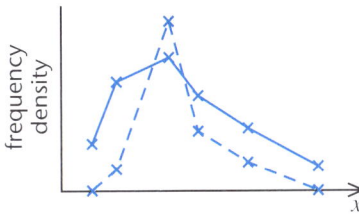

In a **cumulative frequency diagram**, cumulative frequencies (running totals) are plotted against upper class boundaries of the intervals. The median, Q2, and quartiles, Q1 and Q3, can be estimated from the graph. For n items,

Q1 is the value 25% ($\frac{1}{4}n$) through the data,
Q2 is the value 50% ($\frac{1}{2}n$) through the data,
Q3 is the value 75% ($\frac{3}{4}n$) through the data.

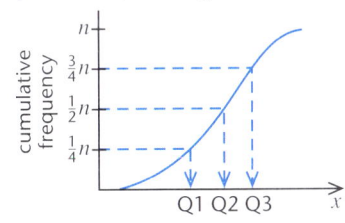

Progress check

1. For the numbers 8, 14, 20, 1, 6, 6, 13, 12, 4,
   (a) find (i) the mean (ii) the standard deviation (iii) the median
   (iv) the interquartile range
   (b) draw a box and whisker plot.

2. Find the mean and standard deviation of this set of data:

x	0	1	2	3
f	4	6	7	3

3.
Mass (g)	$20 \leqslant m < 40$	$40 \leqslant m < 50$	$50 \leqslant m < 55$	$55 \leqslant m < 60$	$60 \leqslant m < 90$
Frequency	10	12	10	7	9

   For the above data
   (a) estimate the mean and the standard deviation
   (b) draw a histogram and state the modal class
   (c) draw a cumulative frequency curve and use it to estimate
   (i) the median (ii) the interquartile range.

1. (a) (i) 9.33 (ii) 5.56 (iii) 8 (iv) 8.5
   (b) [box plot: 0 to 20]
2. 1.45, 0.97
3. (a) 50.89, 14.68
   (b) modal class is $50 \leqslant m < 55$
   (c) (i) 51 g (ii) 16 g

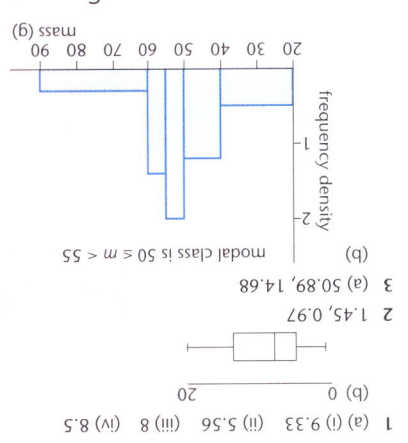

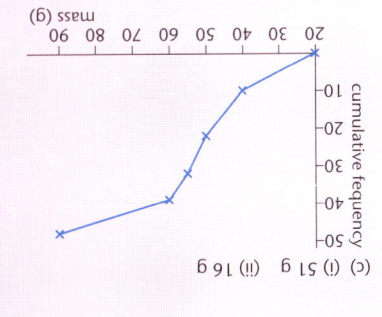

77

Probability and Statistics 1

4.2 Probability

After studying this section you should be able to:

- use Venn diagrams to represent and interpret combined events
- use set notation to express and apply probability laws
- use tree diagrams to represent a problem
- know the formulae for permutation and combinations and recognise when to apply them

Venn diagrams and probability laws

Venn diagrams provide a useful way to represent information about **sets** of objects.

The reason for mentioning them here is that the diagrams, and the notation used to express results, have a direct interpretation and application in probability theory.

In each diagram, the rectangle represents the set S of all objects under consideration.

A ∪ B is read as A union B.

The circles represent particular sets A and B of objects within the set S.

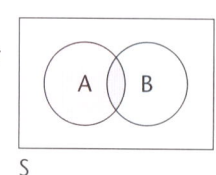

A ∪ B represents the set of all objects that belong to *either* A or B or *both*.

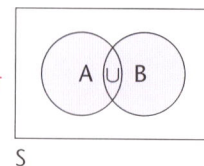

A ∩ B is read as A intersection B.

A ∩ B represents the set of those objects that belong to *both* A *and* B.

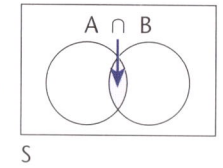

A′ represents the set of objects in S that do not belong to A.

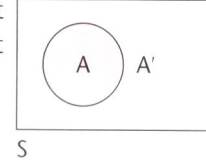

To make use of these ideas in probability theory:

- The objects are interpreted as **outcomes** for a particular situation.
- The set S is the **sample space** of all possible outcomes for the situation.
- The sets A and B are **events** defined by a particular choice of outcomes.
- P(A) for **example**, represents the probability that the event A occurs.
- P(A ∪ B) represents the probability that *either* A or B or *both* occurs.
- P(A ∩ B) represents the probability that *both* A *and* B occur.
- P(A′) represents the probability that A does *not* occur.

The **addition rule** states that

P(A ∪ B) = P(A) + P(B) − P(A ∩ B).

A and B are described as **mutually exclusive events** if they have no outcomes in common. These are events that cannot both occur at the same time.

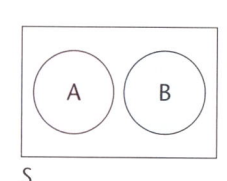

In this case P(A ∩ B) = 0 and the addition law becomes

P(A ∪ B) = P(A) + P(B).

P(A ∩ B) = 0

The events A and A′ are mutually exclusive for any event A.
It follows that P(A ∪ A′) = P(A) + P(A′).
Since one of the events A or A′ must occur, P(A ∪ A′) = 1 so P(A) + P(A′) = 1.

Probability and Statistics 1

This is usually written as $P(A') = 1 - P(A)$.

The **multiplication rule** states that:

$$P(A \cap B) = P(A \mid B) \times P(B),$$

where $P(A \mid B)$ means the probability that A occurs *given that* B has *already* occurred.

$P(A \mid B)$ is described as a **conditional probability**, i.e. it represents the probability of A *conditional* upon B having occurred. Re-arranging the multiplication rule gives

$$P(A \mid B) = \frac{P(A \cap B)}{P(B)}.$$

If A and B are **independent events** then the probability that either event occurs is not affected by whether the other event has already occurred.

In this case, $P(A \mid B) = P(A)$ and the multiplication rule becomes

$$P(A \cap B) = P(A) \times P(B).$$

Example
The events A and B are independent. $P(A) = 0.3$ and $P(B) = 0.6$
Find (a) $P(A \cup B)$ (b) $P(A' \cap B)$.

(a) $P(A \cup B) = P(A) + P(B) - P(A \cap B)$.

> In some problems you need to define the events first before you can apply the probability rules.

Since A and B are independent, $P(A \cap B) = P(A) \times P(B)$.

So $P(A \cup B) = 0.3 + 0.6 - 0.3 \times 0.6 = 0.72$

(b) A and B are independent $\Rightarrow$ A' and B are independent.

> This is true whether or not the events are independent.

So $P(A' \cap B) = P(A') \times P(B) = (1 - 0.3) \times 0.6 = 0.42$

A **tree diagram** is a useful way to represent the probabilities of combined events. Each path through the diagram corresponds to a particular sequence of events and the multiplication rule is used to find its probability. When more than one path satisfies the conditions of a problem, these probabilities are added.

Example
The probability that Chris gets grade A for Maths is 0.6 and the corresponding probability for English is 0.7. The events are independent.

(a) Calculate the probability that Chris gets just one A in the two subjects.
(b) Given that Chris gets just one grade A, find the probability that it is for Maths.

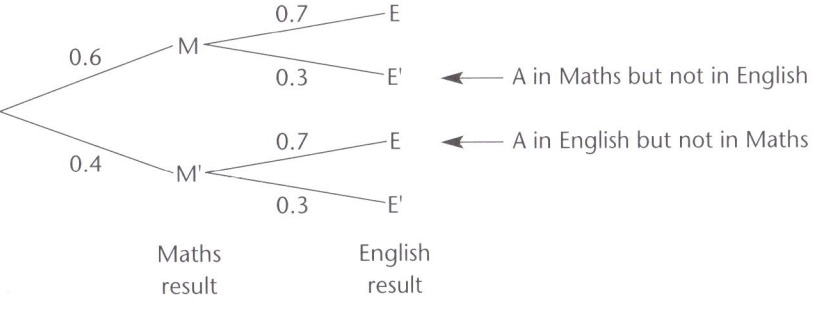

> Show the information on a tree diagram.

```
                0.7      E
         0.6  M
                0.3      E'   ← A in Maths but not in English

                0.7      E    ← A in English but not in Maths
         0.4  M'
                0.3      E'

         Maths         English
         result        result
```

> Multiply the probabilities along the branches.

(a) P(just one A) $= 0.6 \times 0.3 + 0.4 \times 0.7$
$= 0.18 + 0.28$
$= 0.46$

79

Probability and Statistics 1

> Use conditional probability.

(b) P(A in Maths | just one A) = $\dfrac{\text{P(A in Maths and just one A)}}{\text{P(just one A)}}$

$= \dfrac{0.18}{0.46}$

$= 0.39$ (2 d.p.)

Arrangements and permutations, selections and combinations

There are many situations in probability that can be represented by an **arrangement** of objects (often letters) in a particular order. The following results are useful for calculating the number of possible arrangements.

Description	No of arrangements
n different objects in a straight line.	$n!$
n objects in a straight line, where r objects are the same and the rest are different.	$\dfrac{n!}{r!}$
n objects in a straight line, where r objects of one kind are the same, q objects of another kind are the same and the rest are different.	$\dfrac{n!}{q!r!}$

The number of arrangements that can be made by choosing r objects from n is $^nP_r = \dfrac{n!}{(n-r)!}$. These are known as **permutations**.

The number of **selections** (the order makes no difference) that can be made by choosing r objects from n is $^nC_r = \dfrac{n!}{(n-r)!r!}$. These are known as **combinations**.

Progress check

1. The probability that the event B occurs is 0.7. The probability that events A and B both occur is 0.4. What is the probability that A occurs given that B has already occurred?

2. It is given that P(A) = 0.4, P(B) = 0.7 and P(A ∩ B) = 0.2
   Find (a) P(A ∪ B) (b) P(B | A) (c) P(A | B).

3. A bag contains 3 red counters and 7 green counters. Two counters are taken at random from the bag, without replacement.
   (a) By drawing a tree diagram, or otherwise, find the probability that the counters are different colours.
   (b) Given that the counters are different colours, find the probability that the first counter picked is red.

4. Find the number of arrangements of the letters STATISTICS.

5. Find the number of combinations of seven objects chosen from ten.

1 4/7
2 (a) 0.9 (b) 0.5 (c) $\frac{2}{7}$
3 (a) $\frac{7}{15}$ (b) $\frac{1}{2}$
4 $10! \div (3! \times 3! \times 2!) = 50\,400$
5 120

Probability and Statistics 1

4.3 Discrete random variables

After studying this section you should be able to:
- understand the concept of a discrete random variable, its expectation and variance
- understand the binomial and geometric distributions and select the appropriate model for a particular situation

Discrete random variables

The score obtained when a dice is thrown is an example of a **random variable**. Since only specific values may be obtained, it is a **discrete random variable**. In the usual notation:

- A capital letter such as X is used to denote the random variable.
- A lower case letter such as x is used to represent a particular value of X.

This is a useful result.

For a **discrete random variable** X with probability function $P(X = x)$, the sum of the probabilities is 1, i.e. $\sum P(X = x) = 1$.

This is usually shown in a table.

The **probability distribution** of X is given by the set of all possible values of x, together with the values of $P(X = x)$.

Expectation, variance and standard deviation

The symbol μ is used for the **mean value of X**. This is also known as the **expected value, or expectation, of X** and is written $E(X)$.

For any discrete random variable,

$$\mu = E(X) = \sum x P(X = x).$$

The expected value of a function of a random variable can be found in a similar way.

For example, $E(X^2) = \sum x^2 P(X = x)$.

Using σ for the **standard deviation of X** and $\text{Var}(X)$ for the **variance of X**:

$\sigma = \sqrt{\text{Var}(x)}$

$$\sigma^2 = \text{Var}(X) = E(X^2) - [E(X)]^2 = E(X^2) - \mu^2.$$

Example

The table shows the probability distribution for a random variable X.

x	1	2	3
$P(X = x)$	0.2	0.5	0.3

Notice that $\sum P(X = x) = 1$

Calculate:
(a) $P(X > 1)$ (b) $E(X)$ (c) The standard deviation of X.

(a) $P(X > 1) = 0.5 + 0.3 = 0.8$

(b) $E(X) = \sum x P(X = x) = 1 \times 0.2 + 2 \times 0.5 + 3 \times 0.3 = 2.1$

This is the mean of X.

(c) $E(X^2) = \sum x^2 P(X = x) = 1^2 \times 0.2 + 2^2 \times 0.5 + 3^2 \times 0.3 = 4.9$

$\text{Var}(X) = E(X^2) - [E(X)]^2 = 4.9 - 2.1^2 = 0.49$

Standard deviation $\sigma = \sqrt{0.49} = 0.7$

Probability and Statistics 1

The binomial probability distribution

Suppose that n **independent trials** are carried out, each with a **constant probability of success** p and corresponding probability of failure q. Then $p + q = 1$ and the probability of r successes is given by the $(r+1)$th term in the binomial expansion of $(p+q)^n$.

(You need to check that these conditions are satisfied before applying any of the results.)

> Use nCr on the calculator for the value of $\binom{n}{r}$.

Using the random variable X to represent the number of successes in n trials, the probability function is given by $P(X = r) = \binom{n}{r} p^r q^{n-r}$ $r = 0, 1, 2, ..., n$, where $q = 1 - p$.

With this notation, the **binomial probability distribution** may be written as

r:	0	1	2	...	n
$P(X=r)$:	q^n	npq^{n-1}	$\dfrac{n(n-1)}{2}p^2 q^{n-2}$	...	p^n

If X has a binomial distribution with **parameters** n and p then this is written as $X \sim B(n, p)$.

> The standard deviation $\sigma = \sqrt{np(1-p)}$.

In general, if $X \sim B(n, p)$ then $\mu = E(X) = np$ and $Var(X) = np(1-p)$.

Example
The random variable $X \sim B(8, 0.4)$ find:

(a) $P(X = 2)$ (b) $P(X \leq 2)$ (c) $E(X)$ (d) $Var(X)$.

(a) $P(X = 2) = \binom{8}{2}(0.4)^2(0.6)^6 = 0.2090... = 0.209$ to 3 d.p.

> $P(X \leq r)$ may also be found using cumulative Binomial distribution tables. This is particularly useful when r is large.

(b) $P(X \leq 2) = P(X = 0) + P(X = 1) + P(X = 2)$

$= (0.6)^8 + 8(0.4)(0.6)^7 + 0.2090... = 0.3153...$
$= 0.315$ to 3 d.p.

(c) $E(X) = np = 8 \times 0.4 = 3.2$

(d) $Var(X) = np(1-p) = 8 \times 0.4 \times 0.6 = 1.92$

Probability and Statistics 1

The geometric distribution

The **geometric distribution** is used to model the number of independent trials needed before a particular outcome occurs. Taking X to represent this value and p to be the fixed probability that the outcome occurs:

$$P(X = x) = p(1 - p)^{x-1} \quad \text{for} \quad x = 1, 2, 3, \ldots .$$

$$P(X > x) = (1 - p)^x$$

The **parameter** is p.

$$E(X) = \frac{1}{p}$$

The notation Geo(p) may be used to represent this distribution.

Example
A dice is weighted so that the probability it shows a six on any throw is 0.2. The dice is thrown repeatedly. X is the number of throws up to and including the first six.
Find (a) $P(X = 3)$ (b) $P(X < 6)$ (c) $E(X)$

X has the distribution Geo(0.2).

(a) $P(X = 3) = 0.2 \times 0.8^2 = 0.128$

(b) $P(X < 6) = 1 - P(X > 5) = 1 - 0.8^5 = 0.672$ (3 d.p.)

(c) $E(X) = \dfrac{1}{0.2} = 5$

Progress check

1. The random variable X has the probability distribution shown. Find the value of k.

x:	0	1	2
$P(X=x)$:	0.3	0.2	k

2. Find (a) $E(X)$ (b) $Var(X)$ for the probability distribution in question 1.

3. Given that $X \sim B(20, 0.3)$ write down (a) $E(X)$ (b) $Var(X)$.

4. Given that $X \sim$ Geo(0.25) write down $E(X)$.

Answers (inverted):
1. $k = 0.5$
2. (a) 1.2 (b) 0.76
3. (a) 6 (b) 4.2
4. 4

Probability and Statistics 1

4.4 Correlation and regression

After studying this section you should be able to:

- calculate and interpret the product–moment correlation coefficient
- calculate and interpret Spearman's coefficient of rank correlation
- recognise an independent (or controlled) variable and a dependent variable
- find the equations of least squares regression lines
- use a suitable regression line to estimate a value

All of the work in this section relates to the treatment of **bivariate data**. This is data in which each data point is defined by two variables.

Correlation

A **scatter diagram** may be used to represent bivariate data. The extent to which the points approximate to a straight line gives an indication of the strength of a linear relationship between the variables, known as the **linear correlation**.

One way to arrive at a numerical measure of the correlation is to use the **product–moment correlation coefficient**, r.

> You should only use this method when both variables are normally distributed.

For n pairs of (x, y) values:

$$S_{xx} = \Sigma(x - \bar{x})^2 = \Sigma x^2 - \frac{(\Sigma x)^2}{n} \qquad S_{yy} = \Sigma(y - \bar{y})^2 = \Sigma y^2 - \frac{(\Sigma y)^2}{n}$$

$$S_{xy} = \Sigma(x - \bar{x})(y - \bar{y}) = \Sigma xy - \frac{(\Sigma x)(\Sigma y)}{n}$$

and the product–moment correlation coefficient is given by:

> You may be able to obtain the value of r directly from your calculator.

$$r = \frac{S_{xy}}{\sqrt{S_{xx} S_{yy}}}$$

This gives values of r between –1 (representing a perfect negative correlation) and +1 (representing a perfect positive correlation).

An alternative measure of correlation is given by **Spearman's rank correlation coefficient**, r_s.

The set of values for each variable must first be ranked from largest to smallest.

For each data pair, the difference between the ranks of the two variables is denoted by d.

The rank correlation coefficient is given by $r_s = 1 - \dfrac{6\Sigma d^2}{n(n^2 - 1)}$.

This gives values of r_s between –1 (representing a perfect negative correlation between the rankings) and +1 (representing a perfect positive correlation between the rankings).

The two correlation coefficients given above are comparable but not necessarily equal.

Note that the value of a correlation coefficient is unaffected by linear transformations (coding) of the variables.

Probability and Statistics 1

Regression

Whereas correlation is determined by the *strength* of a linear relationship between the two variables, **regression** is about the *form* of the relationship given by the equation of a **regression line**. (Make sure you known the difference between correlation and regression.)

The purpose in establishing the equation of a regression line is to make predictions about the values of one variable (known as the **dependent variable**) for some given values of the other variable (known as the **independent (or controlled) variable**).

> The **dependent** variable is sometimes called the **response** variable and the **independent** variable is sometimes called the **explanatory** variable.

Predictions should only be made for values within the range of readings of the independent (or controlled) variable. **Extrapolation** for values outside this range is unreliable. Another factor affecting the accuracy of any predictions is the influence of **outliers** on the equation of the regression line.

Figure 1 illustrates *y*-residuals given by

$$d = \text{(observed value of } y\text{)} - \text{(predicted value of } y\text{)}.$$

Figure 2 illustrates *x*-residuals given by

$$d = \text{(observed value of } x\text{)} - \text{(predicted value of } x\text{)}.$$

A **least squares regression line** is a line for which the sum of the squares of either the *x*-residuals or *y*-residuals is minimised.

Figure 1

Figure 2

> Both lines always pass through the point $(\bar{x}, \bar{y})$.

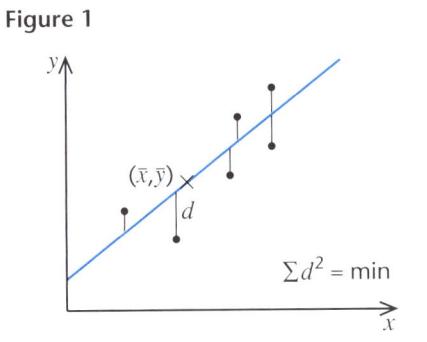

This gives the regression line of y on x as

$$y = a + bx,$$

where $b = \dfrac{S_{xy}}{S_{xx}}$ and $a = \bar{y} - b\bar{x}$.

Use this equation to estimate values of y for given values of x when x is the independent variable and y is the dependent variable.

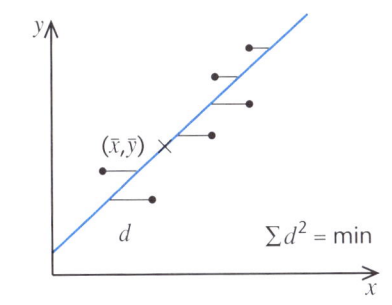

This gives the regression line of x on y as

$$x = c + dy,$$

where $d = \dfrac{S_{xy}}{S_{yy}}$ and $c = \bar{x} - d\bar{y}$.

Use this equation to estimate values of x for given values of y when y is the independent variable and x is the dependent variable.

> The values of a and b may be obtained directly from some calculators.

Unless there is perfect correlation between the variables, the two regression lines will be different and you cannot rearrange one equation to obtain the other.

Progress check

The summary data for 10 pairs of (x, y) values is as follows:

$$\sum x = 146, \quad \sum x^2 = 2208, \quad \sum y = 147, \quad \sum y^2 = 2247, \quad \sum xy = 2211.$$

1. Find the value of the product–moment correlation coefficient between x and y.
2. Find the equation of the least squares regression line of y on x.

1. 0.799
2. $y = 2.317 + 0.848x$

Probability and Statistics 1

Sample questions and model answers

1

Paul swims 20 lengths of a swimming pool in training for a competition. His times, x seconds, for completing lengths of the pool are summarised by:

$$\Sigma x = 372 \qquad \Sigma x^2 = 6952.$$

(a) Find the mean and standard deviation of Paul's times for completing one length of the pool.

(b) A month ago, Paul's mean time per length was 19.8 s with a standard deviation of 1.72 s. Comment on the change in his performance indicated by these results.

Scientific calculators can find the standard deviation from the raw data, but you need to know the formulae because the data may be given in summarised form.

(a) $$\bar{x} = \frac{\Sigma x}{n} = \frac{372}{20}$$

$$= 18.6$$

Paul's mean time per length is 18.6 s.

$$\sqrt{\frac{\Sigma x^2}{n} - \bar{x}^2} = \sqrt{\frac{6952}{20} - 18.6^2}$$

$$= 1.28$$

The standard deviation of Paul's times is 1.28 s.

Quote the formula and substitute the information.

(b) The mean time has been reduced, suggesting an improvement in performance. The standard deviation has also been reduced, which suggests a greater consistency of performance.

You may be expected to interpret results within the context of the question and make comparisons.

Probability and Statistics 1

Sample questions and model answers (continued)

2

The letters of the word ASSOCIATES are jumbled up and arranged in a straight line.

(a) How many different arrangements are there?

(b) How many of the arrangements start with O and end with I?

(c) Find the number of arrangements in which the vowels are grouped together and the probability that this occurs.

(a) There are two As, three Ss and one of each of the other letters.
There are 10 letters altogether.

The number of arrangements is given by

$$\frac{10!}{2!\,3!} = 302\,400$$

> See the formula on page 80 for dealing with arrangements that include repeated letters.

(b) Fixing one letter at each end leaves eight positions to fill.
This includes two As and three Ss as before.

The number of arrangements is given by

$$\frac{8!}{2!\,3!} = 3360.$$

> Filling positions is a useful way to think about the problem.

(c) There are five vowels, AAEIO, that will move together and effectively occupy one position.

These may be arranged in $\frac{5!}{2!} = 60$ ways.

The vowels together with three Ss and one T and one C occupy six positions.

These may be arranged in $\frac{6!}{3!} = 120$ ways.

The total number of arrangements that group the vowels together is $60 \times 120 = 7200$.

The probability that this occurs is $\frac{7200}{302\,400} = \frac{1}{42}$.

> You need to think about the arrangements of vowels within their group.

> You also need to think of the vowels as a single unit or 'letter' to be arranged with the remaining letters.

> To find the total number of arrangements you need to multiply the results together.

Probability and Statistics 1

Practice examination questions

1 The letters in the word COPPER are written on separate pieces of card placed on a table.
A letter is chosen by picking up the card on which it is written.

(a) Find the number of different selections of four letters that can be made.

(b) Find the number of different arrangements of four letters that can be made.

2 A discrete random variable X has the probability function $P(X = x)$ given by

$$P(X = x) = \frac{x^2}{k} \quad \text{for} \quad x = 1, 2, 3.$$

$$P(X = x) = 0 \quad \text{otherwise.}$$

(a) Find the value of k.

(b) Calculate the mean and variance of the probability distribution.

3 X is a random variable such that $X \sim B(20, 0.4)$.

(a) Find the mean and variance of X.

(b) Find the probability that the value of X lies between 2 and 4 inclusive.

(c) Find the probability that X is greater than or equal to 5.

4 A random variable X has the distribution Geo(0.3).
Find:

(a) $P(X = 4)$

(b) $P(X > 2)$

(c) $E(X)$.

Probability and Statistics 1

Practice examination questions (continued)

5 (a) Three cards are drawn at random, without replacement, from a pack of playing cards.
   Calculate the probability that:

   (i)   All three cards are red.
   (ii) All three cards are the same colour.
   (iii) At least one of the cards is red.

   (b) The cards are returned to the pack and the pack is shuffled. One more card is drawn.
   Find the probability that this card is an ace, given that it is not a diamond.

6 Two teachers independently mark a sample of five pieces of students' work as part of a standardisation procedure. The table shows how the marks were allocated.

Sample No	1	2	3	4	5
Teacher A	39	72	51	49	68
Teacher B	46	80	53	54	73

   (a) Calculate Spearman's rank correlation coefficient for the two sets of marks.

   (b) Does a high positive correlation between two such sets of marks mean that the scripts have necessarily been marked to the same standard?
   Explain your answer.

7 A machine distributes 'lucky pendants' to one in four packets of cereal at random. A box contains 20 packets of the cereal.

   (a) Using tables of cumulative binomial probabilities, or otherwise, find the probability that in a randomly selected box:

   (i) At least six packets contain a lucky pendant.
   (ii) Exactly six packets contain a lucky pendant.

   (b) Five boxes are randomly selected. Find the probability that exactly two of the boxes have at least six packets that contain a lucky pendant.

Practice examination questions (continued)

8 Twelve students in the sixth form of a school study both mathematics and physics at A Level.
The scores in an end of term test are shown for 10 of these students.

Maths	48	39	52	55	71	62	76	54	61	58
Physics	41	35	50	47	63	48	70	45	46	39

Using x to represent the score in maths and y to represent the corresponding score in physics:

$$\Sigma x = 576,\ \Sigma y = 484,\ \Sigma x^2 = 34216,\ \Sigma y^2 = 24450,\ \Sigma xy = 28785$$

(a) Calculate the product–moment correlation coefficient, r, for the data.

(b) Find the equation of the regression line of maths scores on physics scores.

(c) Two students took the physics test but missed the maths test.

  (i) Estimate the maths score for one of these students who obtained a physics score of 42.

  (ii) Comment on the validity of estimating the maths score of the second student who obtained a physics score of 80.

Chapter 5
Decision Mathematics 1

The following topics are covered in this chapter:

- Algorithms
- Graphs and networks
- Linear programming

5.1 Algorithms

After studying this section you should be able to:

- implement an algorithm given by text or flow chart
- implement the following algorithms from memory:
  - bubble sort
  - shuttle sort
  - bin packing

Introducing algorithms

Algorithms play a fundamental role throughout Decision Mathematics.

An **algorithm** is a finite sequence of precise instructions used to solve a problem.

One way to define an algorithm is to simply list all of the necessary steps. In more complex situations, the process may be easier to follow if the algorithm is presented as a **flow diagram**.

You need to be able to implement an algorithm presented in either form.

Example
Implement the algorithm given by the flow chart and state what the written values represent.

The table shows how the values of I and J change as the algorithm is implemented.

I	1	2	3	4	5
J	1	3	6	10	15

The written values are 1, 3, 6, 10, 15 and these are the first five triangle numbers.

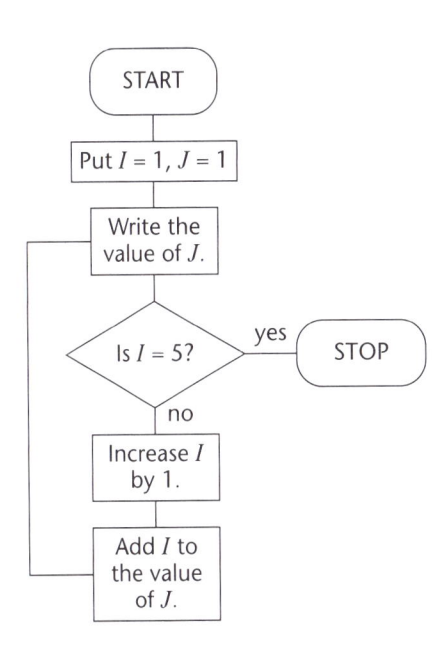

Flow chart:
- START
- Put $I = 1$, $J = 1$
- Write the value of J.
- Is $I = 5$? yes → STOP
- no → Increase I by 1.
- Add I to the value of J.
- (loop back to Write the value of J.)

The algorithm may be expressed in words as:

Write down a sequence of 5 numbers starting with 1 and then increasing, first by 2 and then by 1 more each time from term to term.

This version is more concise but not so easy to understand.

91

Decision Mathematics 1

You need to know and be able to implement the following algorithms.

The bubble sort

As its name suggests, the **bubble sort** is an algorithm for sorting a list in a particular order.

Step 1 Compare the first two items of the list and switch them if they are in the wrong order.

Step 2 Compare the second and third items of the list and, again, switch them if they are in the wrong order.

Step 3 Continue in the same way until you reach the end of the list. This completes the first **pass** through the list.

Step 4 Make repeated passes through the list until a pass produces no change.

Example Sort the list 5, 7, 3, 11, 6, 8 into ascending order.

The first pass gives

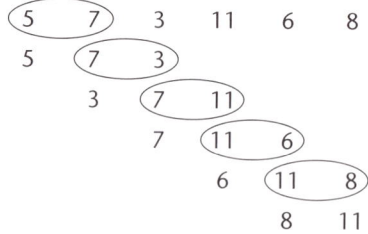

```
 5    7    3   11    6    8
 5    7    3
      3    7   11
           7   11    6
                6   11    8
                     8   11
```

The second pass produces   3, 5, 6, 7, 8, 11.
The third pass produces   3, 5, 6, 7, 8, 11.

The third pass made no change so the ordering is complete.

> Don't just look at the list and say that the numbers are now in order. Use the algorithm to decide when to stop.

The shuttle sort

In the shuttle sort, a pass involves comparing one new pair of items and, if a swap is made, shuttling back to compare previous items as detailed below.

Step 1 In the first pass, compare the first two items in the list and swap if necessary.

Step 2 In the second pass, compare the second and third items and swap if necessary. If a swap is made, compare the first and second items and swap if necessary.

Step 3 In the third pass, compare the third and fourth items and swap if necessary. If a swap is made, shuttle back and compare the second and third items, swapping if necessary. If a swap is made, shuttle back again and compare the first and second items, swapping if necessary.

Step 4 For a list of n items, continue in the same way until the $(n-1)$th pass is complete.

Example Sort the list 5, 7, 3, 11, 6, 8 into ascending order.

First pass	5	7	3	11	6	8
Second pass	5	7	3	11	6	8
	5	3	7	11	6	8
	3	5	7	11	6	8
Third pass	3	5	7	11	6	8
Fourth pass	3	5	7	11	6	8
	3	5	7	6	11	8
	3	5	6	7	11	8
	3	5	6	7	11	8
Fifth pass	3	5	6	7	11	8
	3	5	6	7	8	11

The ordered list is 3, 5, 6, 7, 8, 11.

Decision Mathematics 1

Bin packing

The bin packing algorithm is used to solve problems that can be represented by the need to pack some boxes of equal cross-section but different heights into bins, with the same cross-section as the boxes, using as few bins as possible.

There is no known algorithm that will always provide the *best* solution. You need to be familiar with the two algorithms below that attempt to provide a *good* solution. Such algorithms are known as **heuristic** algorithms.

First-fit algorithm

Take each box in turn from the order given and pack it into the first available bin.

First-fit decreasing algorithm

Step 1 Arrange the boxes in decreasing order of size.
Step 2 Implement the first-fit algorithm starting with the largest box.

Order of an algorithm

The **order** of an algorithm is used to compare the times taken by the algorithm to complete tasks of different lengths. For example, if an algorithm has **linear order** then giving it twice as much work to do will take it twice as long. If, however, the algorithm has **quadratic order** then doubling the amount of work will take the algorithm $2^2 = 4$ times as long.

Example
The bubble sort algorithm has quadratic order. A computer takes 0.2 seconds to sort a list of 400 items using the bubble sort. How long will it take the computer to sort a list of 1200 items using the bubble sort?

$1200 = 400 \times 3$

The computer will take $0.2 \times 3^2 = 1.8$ seconds to sort the 1200 items.

Progress check

You can represent each worker as a 'bin' of 'height' 8 hours.

1. Write the letters P, Q, B, C, A, D in alphabetical order using the bubble sort algorithm.

2. A project involves activities A–H with durations in hours as given in the table.

A	B	C	D	E	F	G	H
3	1	5	4	2	3	4	2

The project is to be completed in 8 hours.
   (a) Use the first-fit algorithm to try to find the minimum number of workers needed.
   (b) Find a better solution.

1 A, B, C, D, P, Q.
2 First-fit: 4 workers; A better solution is (A, B, D) (C, F) (E, G, H).

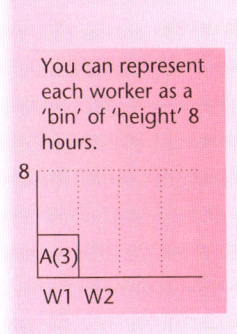

Decision Mathematics 1

5.2 Graphs and networks

LEARNING SUMMARY

After studying this section you should be able to:
- understand the use of terminology associated with graphs and networks
- use Prim's and Kruskal's algorithms for constructing a minimum spanning tree
- use Dijkstra's algorithm to find a shortest path between two vertices
- find upper and lower bounds for the travelling salesperson problem
- use the route inspection algorithm to find a route of minimum weight that traverses every arc of a network

Defining terms

It is worth spending time getting to know all of the terms. Their definitions are quite detailed so you will need to keep reminding yourself of what each term means.

- A **graph** is a set of points, called **vertices** or **nodes**, connected by lines called **edges** or **arcs**.
- A **simple graph** is one that has no loops and in which no pair of vertices are connected by more than one edge.

A simple graph

A non-simple graph — Two edges between A and B

A non-simple graph — A loop around C

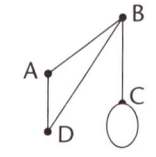

- The number of edges incident on a vertex is called its **order**, **degree** or **valency**. A vertex may be **odd** or **even** depending on whether its order is odd or even.
- A **subgraph** of some graph G is a graph consisting entirely of vertices and edges that belong to G.
- A **directed** edge is an edge that has an associated direction shown by an arrow. A **digraph** is a graph in which the edges are directed.
- A **path** is a finite sequence of edges such that the end vertex of one edge is the start vertex of the next edge. No edge is included more than once.
- A **connected graph** is one in which every pair of vertices is connected by a path.
- A **complete graph** is one in which every pair of vertices is connected by an edge. A complete graph with n vertices is denoted by K_n.
- A **planar graph** is one that can be drawn in a plane such that no two edges meet except at a vertex. K_4 is planar but K_5 is not planar.

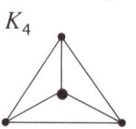

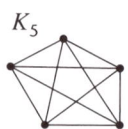

K_4 K_5

- A **cycle** is a path that starts and finishes at the same vertex.
- An **Eulerian** cycle is one that traverses all of the edges of the graph.
- A graph is **Eulerian** if all of its nodes are even.
- A graph is **semi-Eulerian** if it has exactly two odd nodes. Starting at one of the odd nodes, every arc may be traversed once finishing at the other odd node.
- A **tree** is a graph with no cycles.
- A **spanning tree** is a tree whose vertices are all of the vertices of the graph.
- A **network** is a graph that has a number (**weight**) associated with every edge.
- A **minimum spanning tree** (MST) of a network is a spanning tree of minimum possible weight. It is sometimes called a **minimum connector**.

Decision Mathematics 1

Prim's algorithm

Prim's algorithm may be used to find a minimum spanning tree of a network.

Step 1 Choose a starting vertex.

Step 2 Connect it to the vertex that will make an edge of minimum weight.

Step 3 Connect one of the remaining vertices to the tree formed so far, in such a way that the minimum extra weight is added to the tree.

Step 4 Repeat step 3 until all of the vertices are connected.

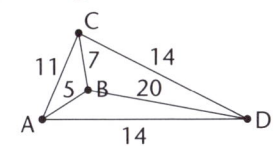

Starting from A gives the MST as

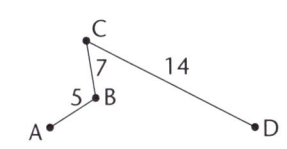

AD could have been used instead of CD

> The number in row A and column B, for example, represents the weight of the edge AB.

The network given above may be represented by the matrix shown.

A version of Prim's algorithm may be used to find the minimum spanning tree directly from a matrix.

	A	B	C	D
A	–	5	–	–
B	5	–	7	–
C	–	7	–	14
D	–	–	14	–

Step 1 Choose a starting vertex. Delete the corresponding row and write 1 above the corresponding column as a label.

Step 2 Circle the smallest undeleted value in the labelled column and delete the row in which it lies.

Step 3 Label the column, corresponding to the vertex of the deleted row, with the next label number.

> If there is more than one smallest value then you can choose which one to circle.

Step 4 Circle the smallest undeleted value of all the values in the labelled columns and delete the row in which it lies.

	1	2	3	
	A	B	C	D
A	–	5	–	–
B	(5)	–	7	–
C	–	(7)	–	14
D	–	–	(14)	–

The circled values correspond to the edges AB, BC, and CD.

Step 5 Repeat steps 3 and 4 until all of the rows are deleted. The circled values then define the edges of the minimum spanning tree.

Kruskal's algorithm

Kruskal's algorithm is an alternative way to find a minimum spanning tree. In this case, the subgraph produced may not be connected until the final stage.

Step 1 Choose an edge with minimum weight as the first subgraph.

Step 2 Find the next edge of minimum weight that will not complete a cycle when taken with the existing subgraph. Include this edge as part of a new subgraph.

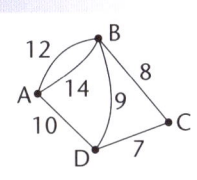

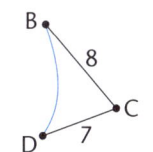

BD cannot be included

Step 3 Repeat step 2 until the subgraph makes a spanning tree.

95

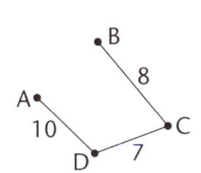

Decision Mathematics 1

Dijkstra's algorithm

> The word *distance* is used here in place of the more general *weight* because in practical applications of Dijkstra's algorithm the weight so often represents a distance.

Dijkstra's algorithm is used to find the shortest distance between a chosen *start* vertex and any other vertex in a network.

The algorithm is designed for use by computers but, when implementing it without the aid of a computer, a system of labelling is required.

O	L
W	

Order of labelling → O
Label (the minimum distance from the start vertex) ← L
Record your 'Working values' here ← W

> Reading the letters inside the box as OWL might help you to remember which part is for which information.

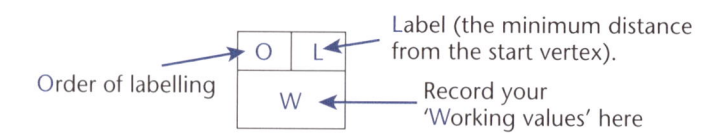

The steps in the algorithm below refer to this system of labelling. Every vertex is labelled in the same way.

Step 1 Give the start vertex a label of 0 and write its order of labelling as 1.

Step 2 For each vertex directly connected to the start vertex, enter the distance from the start vertex as a working value. Enter the smallest working value as a label for that vertex. Record the order in which it was labelled as 2.

Step 3 For each vertex directly connected to the one that was last given a label, add the edge distance onto the label value to obtain a total distance. This distance becomes the working value for the vertex unless a lower one has already been found.

Step 4 Find the vertex with the smallest working value not yet labelled and label it. Record the order in which it was labelled.

Step 5 Repeat steps 3 and 4 until the target vertex is labelled. The value of the label is the minimum distance from the start vertex.

Step 6 To find the path that gives the shortest distance, start at the target vertex and work back towards the start vertex in such a way that an edge is only included if its distance equals the change in the label values.

Example

Use Dijkstra's algorithm to find the shortest distance from A to F in this network and state the route used.

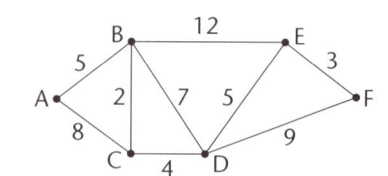

Using Dijkstra's algorithm produces the diagram on the right.

The shortest distance from A to F is 19 (given by the label at F).

Tracing the route backwards using Step 6 gives F, E, D, C, B, A, so the required route is A, B, C, D, E, F.

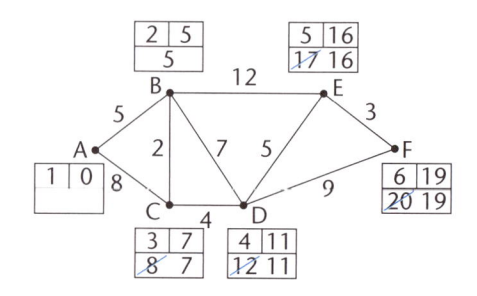

96

Decision Mathematics 1

The travelling salesperson problem

The **travelling salesperson problem** (TSP) is the problem of finding a route of minimum distance that visits every vertex and returns to the start vertex. For a small network it is possible to produce an exhaustive list of all possible routes and choose the one that minimises the total distance. For a large network an exhaustive check is not feasible, even with a computer, because the number of possible routes grows so rapidly.

This is based on the classic situation of a salesperson who wishes to visit a number of towns and return home using the shortest possible route.

It is useful to know within what limits the total distance must lie and there are algorithms that can be used for this purpose.

Finding an upper bound

> **KEY POINT**
>
> An upper bound for the total distance involved in the travelling salesperson problem is given by twice the minimum spanning tree.

The **minimum spanning tree** (MST) for the network ABCD was found earlier.

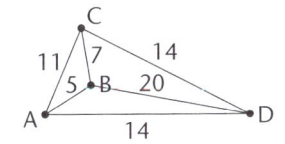

The total length of the MST is $5 + 7 + 14 = 26$.

An upper bound for the TSP is $2 \times 26 = 52$. This corresponds to the route ABCDCBA.

In some cases you might be able to find more than one short-cut that will enable you to reduce the upper bound.

However, an improved (smaller) upper bound can be found by using a **short-cut** from D directly back to A. The route is then ABCDA and the corresponding upper bound is 40.

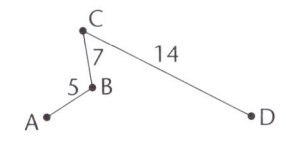

A different approach is to use the **nearest neighbour algorithm**.

Step 1 Choose a starting vertex.

Step 2 Move from your present position to the nearest vertex not yet visited.

Step 3 Repeat step 2 until every vertex has been visited. Return to the start vertex.

For the network ABCD considered above, the nearest neighbour algorithm gives the same result as using the MST with a short-cut. This won't always be the case.

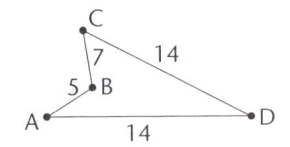

Finding a lower bound

Step 1 Choose a vertex and delete it along with all of the edges connected to it.

Step 2 Find a minimum spanning tree for the remaining part of the network.

Step 3 Add the length of the MST to the lengths of the two shortest deleted edges.

The value found at step 3 is a lower bound for the travelling salesperson problem. It may be possible to find a better (larger) lower bound by choosing to delete a different vertex initially and repeating the process.

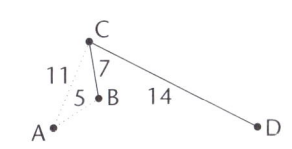

Deleting D to start with gives a lower bound of 40. This corresponds to a cycle ABCDA and so represents the solution of the TSP.

Using the network ABCD again and deleting A gives the minimum spanning tree BCD of length $7 + 14 = 21$. Adding the two shortest deleted lengths gives a lower bound of 37.

Decision Mathematics 1

The route inspection problem

Again, it is common to use length rather than weight in this context.

The **route inspection problem** is to find a route of minimum total length that traverses every edge of the network, at least once, and returns to the start vertex.

This is also known as the Chinese postman problem.

The algorithm for solving this problem is based on the idea of a **traversable** graph. A graph is traversable if it can be drawn in one continuous movement without going over the same edge more than once. If all of the vertices of the graph are even, then any one of them may be used as the starting point and the same point will be returned to at the end of the movement. Such a graph is said to be **Eulerian**. The only other possibility for a traversable graph is that it has exactly two odd vertices. In this situation, one of the vertices is the start point and the other is the finish point. This type of graph is said to be **Semi-Eulerian**.

Leonhard Euler was the most prolific mathematician of all time. One of his many contributions to the subject was to introduce graph theory as a means of solving problems.

The **route inspection algorithm** is:

Step 1   List all of the odd vertices.

Step 2   Form the list into a set of pairs of odd vertices. Find all such sets.

Step 3   Choose a set. For each pair find a path of minimum length that joins them. Find the total length of these paths for the chosen set.

Step 4   Repeat steps 3 until all sets have been considered.

Repeating the edges between the pairs of odd vertices in this way effectively makes all the vertices even and creates a traversable graph.

Step 5   Choose the set that gives the minimum total. Each pair in the set defines an edge that must be repeated in order to solve the problem.

Progress check

1. (a) Use Prim's algorithm to find the total weight of a minimum spanning tree for this network.
   (b) Verify your answer to part (a) using Kruskal's algorithm.

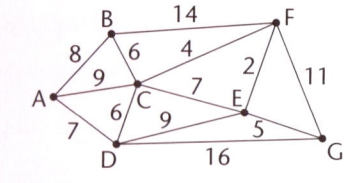

2. Use Dijkstra's algorithm to find the shortest distance from A to G in the network given in question 1. State the route used.

Answers (inverted):
1 (a) 30
2. 20, ACEFG

Decision Mathematics 1

5.3 Linear programming

After studying this section you should be able to:
- formulate a linear programming problem in terms of decision variables
- use a graphical method to represent the constraints and solve the problem
- use the Simplex algorithm to solve the problem algebraically

Formulating a linear programming problem

To formulate a linear programming problem you need to:

- Identify the **variables** in the problem and give each one a label.
- Express the **constraints** of the problem in terms of the variables. You need to include non-negativity constraints such as $x \geq 0$, $y \geq 0$.
- Express the quantity to be optimised in terms of the variables. The expression produced is called the **objective function**.

x and y are often used for the variables.

Typically, this may be to maximise a profit or minimise a loss.

Example

A small company produces two types of armchair. The cost of labour and materials for the two types is shown in the table.

	Labour	Materials
Standard	£30	£25
Deluxe	£40	£50

The total spent on labour must not be more than £1150 and the total spent on materials must not be more than £1250. The profit on a standard chair is £70 and the profit on a deluxe chair is £100. How many chairs of each type should be made to maximise the profit?

In this case, the variables are the number of chairs of each type that may be produced. Using x to represent the number of standard chairs and y to represent the number of deluxe chairs, the constraints may be written as:

$$30x + 40y \leq 1150 \Rightarrow 3x + 4y \leq 115$$
$$25x + 50y \leq 1250 \Rightarrow x + 2y \leq 50$$

It's a good idea to simplify the constraints where possible.

and $x \geq 0$, $y \geq 0$.

Using P to stand for the profit, the problem is to maximise $P = 70x + 100y$.

The graphical method of solution

Each constraint is represented by a region on the graph. It's a good idea to shade out the *unwanted* region for each one. The part that remains unshaded then defines the **feasible region** containing the points that satisfy all of the constraints.

The blue line represents the points where the profit takes a particular value. Moving the line in the direction of the arrow corresponds to increasing the profit. This suggests that the maximum profit occurs at the point X.

X does not represent the solution in this case because both x and y must be integers.

Solving $3x + 4y = 115$ and $x + 2y = 50$ simultaneously gives X as $(15, 17.5)$.

The nearest points with integer coordinates in the feasible region are $(15, 17)$ and $(14, 18)$. The profit, given by $P = 70x + 100y$, is greater at $(14, 18)$.

The maximum profit is made by producing 14 standard and 18 deluxe chairs.

Decision Mathematics 1

The Simplex algorithm

When there are three or more variables, a different approach is needed. The **Simplex algorithm** may be used to solve the problem algebraically. The information must be expressed in the right form before the algorithm can be used.

- First write the constraints, other than the non-negativity conditions, in the form $ax + by + cz \leqslant d$.
- Write the objective function in a form which is to be maximised.
- Add **slack variables** to convert the inequalities into equations.
- Write the information in a table called the **initial tableau**.

> The objective function is already in a form which is to be maximised.
>
> The constraints are in the right form.

Example
Find the maximum value of $P = 2x + 3y + 4z$ subject to the constraints:

$$3x + 2y + z \leqslant 10 \quad \text{[i]}$$
$$2x + 5y + 3z \leqslant 15 \quad \text{[ii]}$$
$$x \geqslant 0, \quad y \geqslant 0, \quad z \geqslant 0.$$

Using slack variables s and t, the inequalities [i] and [ii] become:

$$3x + 2y + z + s = 10$$
$$2x + 5y + 3z + t = 15.$$

The objective function must be rearranged so that all of the terms are on one side of the equation to give:

$$P - 2x - 3y - 4z = 0.$$

The information may now be put into the initial tableau.

	P	x	y	z	s	t	value
The top row shows the objective function	1	−2	−3	−4	0	0	0
This row shows the first constraint	0	3	2	1	1	0	10
This row shows the second constraint	0	2	5	3	0	1	15

The columns for P, s and t contain zeros in every row apart from one. In each case, the remaining row contains 1 and the value of the variable is given in the end column of that row. The value of every other variable is taken to be zero. So, at this stage, the tableau shows that $P = 0$, $s = 10$, $t = 15$ and x, y and z are all zero. This corresponds to the situation at the origin.

> You need to know how to read values from the tableau.

Using the algorithm is equivalent to visiting each vertex of the feasible region until an optimum solution is found. This occurs when there are no negative values in the objective row.

When the present tableau is not optimal, a new tableau is formed as follows:

- The column containing the most negative value in the objective row becomes the **pivotal column**. In this case the pivotal column corresponds to the variable z.
- Now divide each entry in the *value* column by the corresponding entry in the pivotal column provided that the pivotal column entry is positive.
  For the example above this gives $\frac{10}{1} = 10$ and $\frac{15}{3} = 5$
  The smallest of these results relates to the bottom row which is now taken to be the **pivotal row**. The entry lying in both the pivotal column and the pivotal row then becomes the pivot. In this case, the pivot is 3.

Decision Mathematics 1

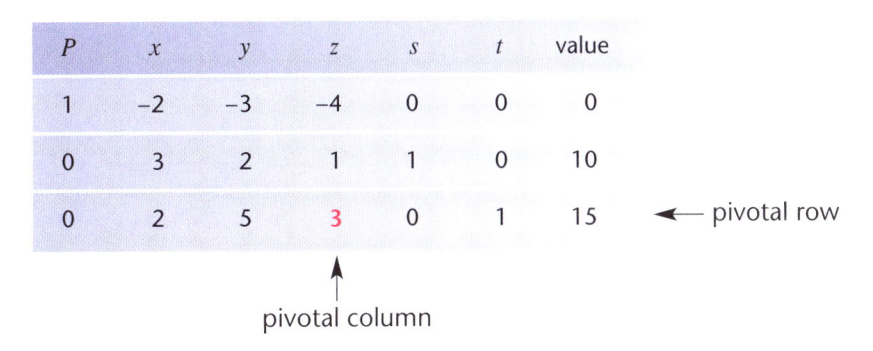

P	x	y	z	s	t	value
1	-2	-3	-4	0	0	0
0	3	2	1	1	0	10
0	2	5	**3**	0	1	15	← pivotal row

↑ pivotal column

- Divide every value in the pivotal row by the pivot, this makes the value of the pivot 1. The convention is to use fraction notation rather than decimals.

P	x	y	z	s	t	value
1	-2	-3	-4	0	0	0
0	3	2	1	1	0	10
0	$\frac{2}{3}$	$\frac{5}{3}$	1	0	$\frac{1}{3}$	5	← pivotal row

- Now turn the other values in the pivotal column into zeros by adding or subtracting multiples of the pivotal row.

P	x	y	z	s	t	value
1	$\frac{2}{3}$	$\frac{11}{3}$	0	0	$\frac{4}{3}$	20
0	$\frac{7}{3}$	$\frac{1}{3}$	0	1	$-\frac{1}{3}$	5
0	$\frac{2}{3}$	$\frac{5}{3}$	1	0	$\frac{1}{3}$	5	← pivotal row

See paragraph on preceding page about reading values from the tableau.

- If there are no negative values in the objective row then the **optimum tableau** has been found. Otherwise choose a new pivotal column and repeat the process.

It is often necessary to repeat the process in order to find the optimum tableau.

Each time through the process is called an **iteration**.

In this case the optimum tableau has been found after just one iteration. It shows that the maximum value of P is 20 and that this occurs when

$x = 0$, $y = 0$, $z = 5$, $s = 5$ and $t = 0$.

Progress check

An initial simplex tableau for a linear programming problem is given by:

P	x	y	z	s	t	value
1	-3	-2	-1	0	0	0
0	2	3	1	1	0	10
0	3	4	2	0	1	18

1. Carry out one iteration to produce the next tableau.

2. Explain why the tableau found is optimal. Write down the maximum value of P and the corresponding values of the other variables.

Answers (rotated):

2. There are no negative values in the objective row.
   $P = 15$, $x = 5$, $y = 0$, $z = 0$, $s = 0$, $t = 3$.

P	x	y	z	s	t	value
1	0	$\frac{5}{2}$	$\frac{1}{2}$	$\frac{3}{2}$	0	15
0	1	$\frac{3}{2}$	$\frac{1}{2}$	$\frac{1}{2}$	0	5
0	0	$-\frac{1}{2}$	$\frac{1}{2}$	$-\frac{3}{2}$	1	3

101

Decision Mathematics 1

Sample questions and model answers

1

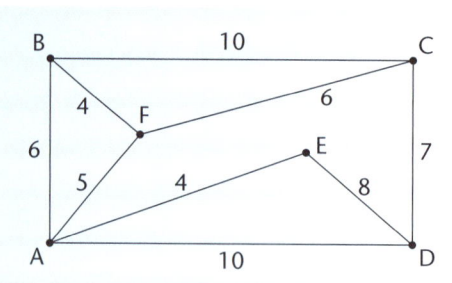

[Graph with vertices A, B, C, D, E, F. Edges: AB=6, BC=10, CD=7, AD=10, BF=4, AF=5, FE=4, CE=6, ED=8]

The diagram represents the system of pathways that a security guard must patrol during his course of duty. The weights on the edges represent the time taken, in minutes, to patrol each pathway.

Find the minimum time required to patrol every pathway at least once and give a possible route.

> *You need to recognise that this question requires the use of the route inspection algorithm.*

> *Step 1 of the algorithm is to list the odd vertices.*

Vertex	Order
A	4
B	3
C	3
D	3
E	2
F	3

The odd vertices are B, C, D and F.

The sets of pairs of odd vertices are: {BC, DF}
{BD, CF}
{BF, CD}.

> *Step 2 of the algorithm is to find the sets of pairs of odd vertices. Each odd vertex appears once in each set.*

For the set {BC, DF}

BC = 10. The shortest route from D to F is DC + CF = 13.

> *There is an edge connecting B and C.*

Total length of the extra paths for this set is
10 + 13 = 23.

> *Step 3 of the algorithm is to find the extra length introduced for each set.*

For the set {BD, CF}

The shortest route from B to D is BA + AD = 16. CF = 6.

Total length of the extra paths for this set is 16 + 6 = 22.

For the set {BF, CD}

> *There is an edge connecting C and F.*

BF = 4 and CD = 7.

Total length of the extra paths for this set is 4 + 7 = 11.

The minimum total corresponds to the set {BF, CD} which means that the edges BF and CD will be repeated.

> *Step 4 of the algorithm is to keep going until all the sets have been considered.*

> *There are many possible routes.*

A possible route is ABFCDEAFBCDA.

The time taken will be 60 + 11 = 71 minutes.

Decision Mathematics 1

Sample questions and model answers (continued)

2

A new theme park has its major attractions at A, B, C, D, E, F and G as shown in the network below. The edges of the network show the *possible* routes of paths connecting the attractions.

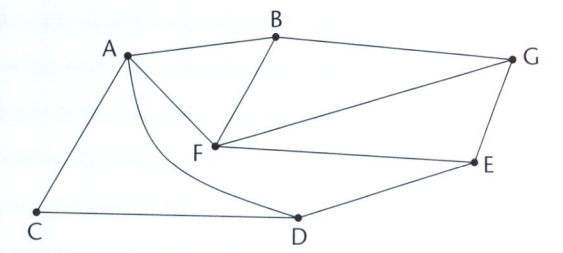

The cost of laying each of these paths, in hundreds of pounds, is given in the table:

	A	B	C	D	E	F	G
A	–	50	40	61	–	38	–
B	50	–	–	–	–	47	54
C	40	–	–	35	–	–	–
D	61	–	35	–	42	–	–
E	–	–	–	42	–	25	10
F	38	47	–	–	25	–	63
G	–	54	–	–	10	63	–

The paths that are actually laid must form a connected network.

Use Prim's algorithm, starting by deleting row A, to find the minimum cost of laying the necessary paths.

Applying the matrix form of Prim's algorithm gives:

	1	7	5	6	3	2	4
	A	B	C	D	E	F	G
~~A~~	–	50	40	61	–	38	–
~~B~~	50	–	–	–	–	(47)	54
~~C~~	(40)	–	–	35	–	–	–
~~D~~	61	–	(35)	–	42	–	–
~~E~~	–	–	–	42	–	(25)	10
~~F~~	(38)	47	–	–	25	–	63
~~G~~	–	54	–	–	(10)	63	–

The minimum cost is found by adding the circled figures and multiplying by £100.

Minimum cost = £19 500.

Decision Mathematics 1

Practice examination questions

1 The table shows the amount of memory taken up by some files on a computer system.

File	A	B	C	D	E	F	G
Memory (Kb)	580	468	610	532	840	590	900

The files are to be transferred onto disks with a capacity of 1.4 Mb (1000 Kb = 1 Mb).

(a) Show that a minimum of four disks is required.

(b) Use the first-fit algorithm to allocate the files to disks.

(c) Use the first-fit decreasing algorithm to show how all of the files may be stored using four disks.

2

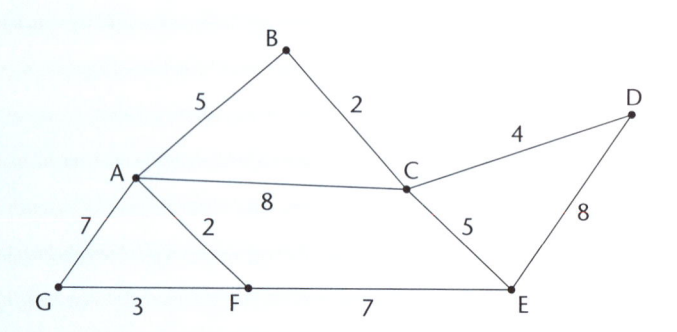

(a) Use Dijkstra's algorithm to find the shortest distance from G to D.

(b) Describe the route taken.

3

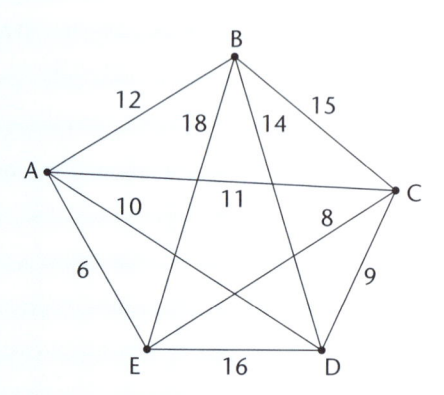

(a) Use Kruskal's algorithm to find the length of the minimum spanning tree for this network.

State the order in which you include each of the edges.

(b) Find an upper bound for the travelling salesperson problem using the minimum spanning tree with shortcuts.

Decision Mathematics 1

Practice examination questions (continued)

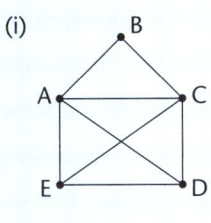

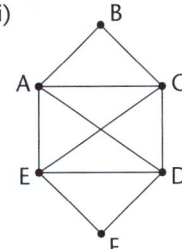

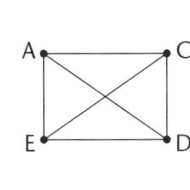

4 (i), (ii), (iii) — three graphs on vertices A, B, C, D, E (and F in (ii))

(a) List the order of the vertices for each graph.

(b) Describe each of the graphs as Eulerian, Semi-Eulerian or neither.

5 A distribution manager has the task of transporting pallets of frozen food between two storage centres. The maximum number of pallets is to be delivered within a daily budget of £2000. Three types of van are available and each van can only do the trip once per day.

The details of what the vans can carry and the daily costs are shown in the table.

At most 10 drivers may be used in one day.

Van type	No. of pallets	Cost/day
Class A	4	£120
Class B	9	£300
Class C	12	£400

Let a represent the number of Class A vans used, b the number of Class B vans used, c the number of Class C vans used and P the total number of pallets transported in a day.

(a) Write an expression for the objective function.

(b) Formulate the task as a linear programming problem.

(c) Describe any special condition that the solution must satisfy.

Decision Mathematics 1

Practice examination questions (continued)

6 A linear programming problem is formulated as:

Maximise $P = x + 2y + 3z$,

subject to $\quad 2x + 6y + z \leqslant 10$

$\quad\quad\quad\quad\quad x + 4y + 5z \leqslant 16$

and $\quad\quad x \geqslant 0, y \geqslant 0, z \geqslant 0$.

(a) Set up an initial Simplex tableau to represent the problem.

(b) Use the Simplex algorithm to solve the problem.
State the corresponding values of x, y and z.

(c) Explain how you know that the optimum solution has been found.

Practice examination answers

Core 1

1 $49^{-\frac{1}{2}} = \frac{1}{49^{\frac{1}{2}}} = \frac{1}{\sqrt{49}} = \frac{1}{7}$.

2 $8^{x-3} = 4^{x+1} \Rightarrow (2^3)^{x-3} = (2^2)^{x+1}$
$\Rightarrow 2^{3(x-3)} = 2^{2(x+1)}$
$\Rightarrow 3(x-3) = 2(x+1)$
$\Rightarrow x = 11$.

3 (a) $x^2 - 4x + 1 = (x-2)^2 - 3$.

(b) $(x-2)^2 - 3 = 0$
$\Rightarrow x - 2 = \pm\sqrt{3}$
$\Rightarrow x = 2 \pm \sqrt{3}$.

(c) $(2, -3)$.

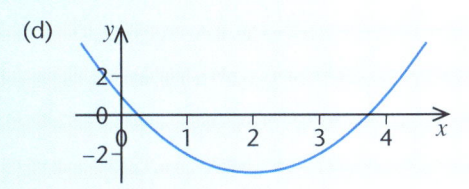

(d) [graph of parabola with vertex at (2, -3), crossing x-axis near 1 and 4, y-intercept at 2]

4 (a) $f'(x) = 2x - x^{-2}$
$f'(2) = 4 - \frac{1}{4} = 3\frac{3}{4}$

(b) $f''(x) = 2 + 2x^{-3}$
$f''(-1) = 2 - 2 = 0$.

5 The coordinates at the points of intersection are given by the simultaneous solution of:

$y = 2x - 2$ [1]

$y = x^2 - x - 6$ [2]

Eliminating y gives
$x^2 - x - 6 = 2x - 2$
$\Rightarrow x^2 - 3x - 4 = 0$
$\Rightarrow (x + 1)(x - 4) = 0$
$\Rightarrow x = -1$ or $x = 4$.

When $x = -1$, $y = -4$.
When $x = 4$, $y = 6$.

The points of intersection are A$(-1, -4)$ and B$(4, 6)$.

6 $3x^2 - 8x - 7 < 2x^2 - 3x - 11$
$\Rightarrow x^2 - 5x + 4 < 0$
$\Rightarrow (x - 1)(x - 4) < 0$
$\Rightarrow 1 < x < 4$.

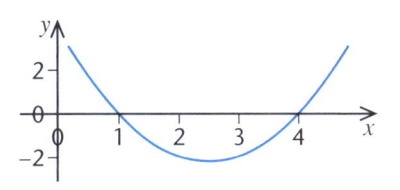

[graph of parabola crossing x-axis at 1 and 4, y-intercept at 2, minimum below x-axis between]

7 (a) Gradient of AB = 6/5.
Using $y - y_1 = m(x - x_1)$ gives
$y - 1 = 6/5(x - 4)$
$\Rightarrow 5y - 5 = 6x - 24$
$\Rightarrow 6x - 5y - 19 = 0$.

(b) The midpoint of A and B is $(1.5, -2)$.
The gradient of the line l is $-5/6$.
Using $y - y_1 = m(x - x_1)$ again gives
$y + 2 = -5/6(x - 1.5)$
$\Rightarrow 6y + 12 = -5x + 7.5$
$\Rightarrow 12y + 24 = -10x + 15$
$\Rightarrow 10x + 12y + 9 = 0$.

(c) At C, $x = 0 \Rightarrow y = -0.75$ so the coordinates of C are $(0, -0.75)$.

8 (a) The equation of the line m is $3x + 2y = 6$
$\Rightarrow y = -3/2 x + 3$
$\Rightarrow$ The gradient of the line m is $-3/2$
$\Rightarrow$ The gradient of the line l is $2/3$.

Using $y - y_1 = m(x - x_1)$ gives
$y - 4 = 2/3(x - 5)$
$\Rightarrow 3y - 12 = 2x - 10$
$\Rightarrow 2x - 3y + 2 = 0$ which is the equation of line l.

(b) At A, $y = 0 \Rightarrow x = -1$.
B lies on the line with equation $3x + 2y = 6$.
At B, $y = 0 \Rightarrow x = 2$.

So the distance AB is given by $2 - (-1) = 3$.

107

Practice examination answers

Core 1 (continued)

9 (a) $y = 2x^3 - 15x^2 - 36x + 10$

$\dfrac{dy}{dx} = 6x^2 - 30x - 36$

At a stationary point

$\dfrac{dy}{dx} = 0 \Rightarrow 6x^2 - 30x - 36 = 0$

$\Rightarrow x^2 - 5x - 6 = 0$

$\Rightarrow (x+1)(x-6) = 0$

$\Rightarrow x = -1 \text{ or } x = 6.$

When $x = -1$, $y = 29$.
When $x = 6$, $y = -314$.
The stationary points are $(-1, 29)$ and $(6, -314)$.

(b) $\dfrac{d^2y}{dx^2} = 12x - 30$

When $x = -1$, $\dfrac{d^2y}{dx^2} = -42 < 0$ maximum.

When $x = 6$, $\dfrac{d^2y}{dx^2} = 42 > 0$ minimum.

(c) The function is decreasing at all of the points between the turning points, i.e. when $-1 < x < 6$.

10 (a) $x + y = 7$.

(b) At P and Q $\quad x + y = 7$ [1]
and $\quad y = (x-1)^2 + 4$ [2]

Substituting for y in [2] gives
$7 - x = x^2 - 2x + 5$
$\Rightarrow x^2 - x - 2 = 0$
$\Rightarrow (x-2)(x+1) = 0$
$\Rightarrow x = 2 \text{ or } x = -1.$
$x = 2 \Rightarrow y = 5$ and $x = -1 \Rightarrow y = 8$.
P is the point $(-1, 8)$ and Q is the point $(2, 5)$.

11 (a) $2x^{\frac{3}{2}}\left(1 - \dfrac{x}{5}\right)$

(b) $(0,0), (5,0)$

(c) $\dfrac{dy}{dx} = 3x^{\frac{1}{2}} - x^{\frac{3}{2}}$

(d) $\dfrac{dy}{dx} = 0 \Rightarrow x^{\frac{1}{2}}(3 - x) = 0$

$\Rightarrow x = 0 \text{ or } x = 3$
Since $x > 0$, $x = 3$
when $x = 3$, $y = 4.157$ (4 s.f.)

$\dfrac{d^2y}{dx^2} = \tfrac{3}{2}x^{-\frac{1}{2}} - \tfrac{3}{2}x^{\frac{1}{2}}$

when $x = 3$, $\dfrac{d^2y}{dx^2} = \tfrac{3}{2}(3^{-\frac{1}{2}} - 3^{\frac{1}{2}}) < 0$

So the turning point at $(3, 4.157)$ is a maximum.

Core 2

1 10th term $= a + 9d = 74$. [1]
Sum of first 20 terms $= 10(2a + 19d) = 1510$

$\Rightarrow \quad 2a + 19d = 151$ [2]

$2 \times$ [1] gives $\quad 2a + 18d = 148$ [3]

[2] – [3] gives $\quad d = 3$

Subs for d in [1] gives $a + 27 = 74 \Rightarrow a = 47$.
First term $= 47$, common difference $= 3$.

2 (a) First term $a = 3$, common ratio $r = 2$.
The 8th term is $ar^7 = 3 \times 2^7 = 384$.

(b) First term $a = 25$, common difference $d = 1.2$
number of terms $n = 50$.

Using $S_n = \dfrac{n}{2}(2a + (n-1)d)$ gives

Sum $= \tfrac{50}{2}(2 \times 25 + 49 \times 1.2) = 2720$.

3 (a) By the factor theorem $f(-2) = 0$
$\Rightarrow (-2)^3 - 4(-2)^2 - 3(-2) + k = 0$
$\Rightarrow -8 - 16 + 6 + k = 0$
$\Rightarrow k = 18$.

(b) $f(x) = x^3 - 4x^2 - 3x + 18$
$\equiv (x+2)(Ax^2 + Bx + C)$.
Comparing coefficients of x^3 gives $A = 1$.
Comparing the constant terms gives $C = 9$.
When $x = -1$
$-1 - 4 + 3 + 18 = A - B + C$
$\Rightarrow 16 = 1 - B + 9 \Rightarrow B = -6$.
$f(x) = (x+2)(x^2 - 6x + 9)$
$= (x+2)(x-3)^2$.

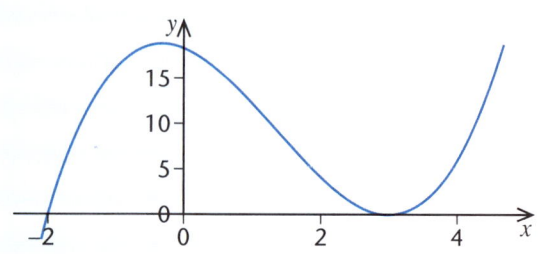

Practice examination answers

Core 2 (continued)

(c) From the sketch, $f(x) \geq 0 \Rightarrow x \geq -2$.

(d) By the remainder theorem, the remainder is given by $f(-1) = 16$.

4 (a) $3\log_a x - 2\log_a y + \log_a(x+1)$
$= \log_a x^3 - \log_a y^2 + \log_a(x+1)$
$= \log_a\left(\dfrac{x^3(x+1)}{y^2}\right)$.

(b) $5^x = 100$
$\Rightarrow \log_{10} 5^x = \log_{10} 100$
$x \log_{10} 5 = \log_{10} 100$
$x = \dfrac{\log_{10} 100}{\log_{10} 5} = 2.861$ to 3 d.p.

5 $\cos^{-1}(0.4) = 66.42\ldots°$
$0° \leq x \leq 360° \Rightarrow 30° \leq 2x + 30 \leq 750°$
Sketch the graph of $y = \cos X$, $30° \leq X \leq 750°$, and the graph of $y = 0.4$

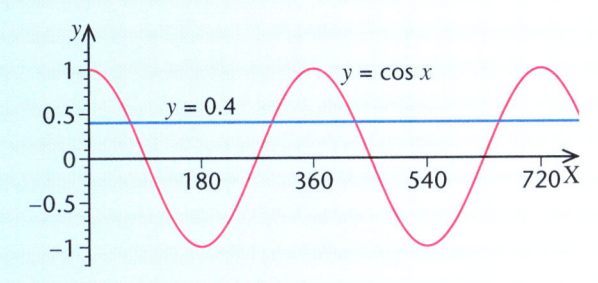

From the sketch, the values of X (i.e $2x + 30$) are:
$66.42\ldots°, 360 - 66.42\ldots°, 360 + 66.42\ldots°,$
$720 - 66.42\ldots°,$
This gives
$x = 18.2°$ or $x = 131.8°$ or $x = 198.2°$ or $x = 311.8°$

6 (a) $\cos x + 3 \sin x \tan x - 2 = 0$
$\Rightarrow \cos^2 x + 3\sin^2 x - 2\cos x = 0$
$\Rightarrow \cos^2 x + 3(1 - \cos^2 x) - 2\cos x = 0$
$\Rightarrow 3 - 2\cos^2 x - 2\cos x = 0$
$\Rightarrow 2\cos^2 x + 2\cos x - 3 = 0$

(b) $\cos x = \dfrac{-2 \pm \sqrt{4 + 24}}{4}$
$\Rightarrow \cos x = 0.8228\ldots$ or $\cos x = -1.8228\ldots$
(No solutions)
$\Rightarrow x = 34.6°$ or $x = 325.4°$

7 (a) So, $f(x) = (x-2)(x^2 + 6x + 9)$
$= (x-2)(x+3)^2$.

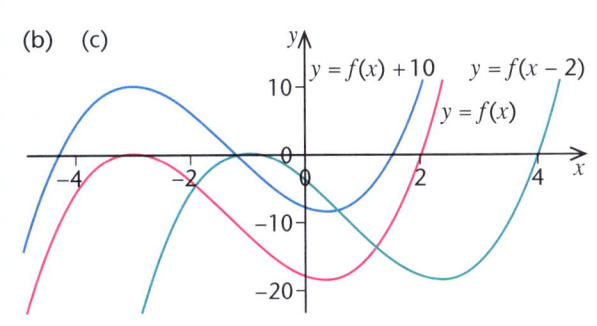

(b) (c)

8 Area $= \displaystyle\int_1^4 3x^{\frac{1}{2}} \, dx$
$= [2x^{\frac{3}{2}}]_1^4 = 2(8-1)$
$= 14.$

9 Area sector = area triangle + area segment
$\Rightarrow \dfrac{1}{2}\theta = \dfrac{1}{2}\sin\theta + \dfrac{\pi}{5}$
$\Rightarrow \theta = \sin\theta + \dfrac{2\pi}{5}.$

10 Area $= \displaystyle\int_2^3 \sqrt{x^2 - 3} \, dx$.

$h = \dfrac{3-2}{5} = 0.2$

$A = 0.1\{(1 + 2.4495) + 2(1.3565 + 1.6613 + 1.9591 + 2.2)\}$
$A = 1.78$ to 2 d.p.

Practice examination answers

Mechanics 1

1 (a) $u = 10$
           $a = -4$
Using $v = u + at$ $v = 10 - 4 \times 5 = -10$.
The speed of the particle when $t = 5$ is 10 m s^{-1}.

(b) $v = 0 \Rightarrow 10 - 4t = 0$
           $\Rightarrow t = 2.5$
Using $s = ut + \tfrac{1}{2}at^2$
when $t = 2.5$ $s = 10 \times 2.5 - 2 \times 6.25$
           $s = 12.5$
when $t = 4$ $s = 10 \times 4 - 2 \times 16$
           $s = 8$.

The total distance travelled between $t = 0$ and $t = 4$ is 12.5 m + 4.5 m = 17 m.

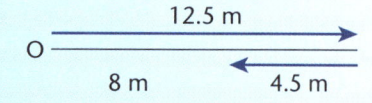

2 (a) The acceleration is positive between $t = 0$ and $t = 25$ so the greatest velocity occurs when $t = 25$.

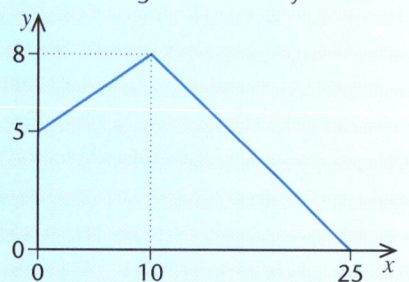

(b) [graph: velocity-time graph with points at (0,5), (10,8), (25,0)]

The area under the graph represents the increase in velocity between $t = 0$ and $t = 25$.
Area = $\tfrac{1}{2} \times 10(5 + 8) + \tfrac{1}{2} \times 15 \times 8 = 125$.
The increase in velocity is 125 m s^{-1}.
The initial velocity is 20 m s^{-1}.
The maximum value of the velocity is 145 m s^{-1}.

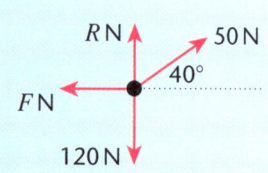

3 [diagram: forces R N upward, 50 N at 40°, F N left, 120 N down]

(a) Resolving vertically
     $R + 50 \sin 40° - 120 = 0$
     $\Rightarrow R = 120 - 50 \sin 40°$
     $\Rightarrow R = 87.86 \ldots$
The normal reaction is 87.9 N to 1 d.p.

(b) Resolving horizontally
     $F - 50 \cos 40° = 0$
     $\Rightarrow F = 38.30 \ldots$
For limiting equilibrium
     $F = \mu R$
giving $38.30\ldots = \mu \times 87.86\ldots$
         $\mu = 0.4359 \ldots$

The coefficient of friction is 0.436 to 3 d.p.

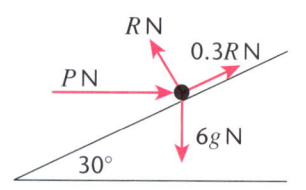

4 [diagram: inclined plane at 30°, forces R N, $0.3R$ N, P N, $6g$ N]

The equations are simpler to deal with by resolving vertically and horizontally, in this case, instead of parallel and normal to the plane.

The minimum value of P corresponds to the case where friction is limiting, as shown in the diagram.

Resolving vertically
     $R \cos 30° + 0.3 R \sin 30° = 6 \times 9.8$

$$R = \frac{6 \times 9.8}{\cos 30° + 0.3 \sin 30°}$$

     $R = 57.87 \ldots$

Resolving horizontally
     $P + 0.3R \cos 30° - R \sin 30° = 0$
     $P = R \sin 30° - 0.3R \cos 30°$
     $P = 13.90 \ldots$

The minimum value of P is 13.9 to 1 d.p.

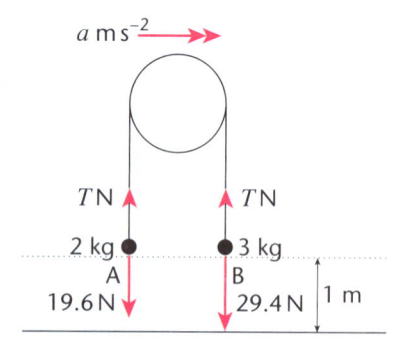

5 [diagram: pulley system with 2 kg (A, 19.6 N) and 3 kg (B, 29.4 N), tensions T N, acceleration a m s^{-2}, height 1 m]

(a) The equations of motion for the particles are:
For A $T - 19.6 = 2a$
For B $29.4 - T = 3a$.
Adding gives $9.8 = 5a \Rightarrow a = 1.96$
The acceleration of the system is 1.96 m s^{-2}

(b) $T - 19.6 = 3.92 \Rightarrow T = 23.52$
The tension in the string is 23.52 N.

(c) For B $u = 0$
          $a = 1.96$,
          $s = 1$.
Using $s = ut + \tfrac{1}{2}at^2$
          $1 = 0 + 0.98t^2$ ($t > 0$)
          $\Rightarrow t = 1.01 \ldots$

Particle B hits the ground after 1.01 s to 2 d.p.

Practice examination answers

Mechanics 1 (continued)

(d) Using $v = u + at$
$v = 0 + 1.96 \times 1.01... = 1.979....$
Particle B hits the ground with speed 1.98 m s^{-1} to 2 d.p.

(e) Particle A rises to height 2 m then the string becomes slack and it behaves as a projectile.
Using $\quad v^2 = u^2 + 2as$
at max height $\quad 0 = 1.979^2 - 19.6s$
$\Rightarrow s = 0.2$
The max height reached by particle A is 2.2 m.

(f) In the extreme situation where the pulley does not move, each particle would be held in equilibrium and so the tension on each side would match the weight of the corresponding particle.
In the given situation, where the pulley is not light or frictionless, it cannot be assumed that the tension on each side of the pulley is the same. This introduces an extra unknown value into the equations. For example, the tension on one side might be labelled T_1 and on the other T_2.

6 (a) $x = t^3 - 6t^2 + 3$
$\dot{x} = 3t^2 - 12t = 3t(t - 4)$
When $t = 0$, $\dot{x} = 0$ so initially at rest. At rest again when $t = 4$
$4^3 - 6 \times 4^2 + 3 = -29$
Max distance in negative direction is 29 m.

(b) The particle is at its initial position when
$t^3 - 6t^2 = 0 \Rightarrow t^2(t - 6) = 0$
Particle passes through initial position when $t = 6$.
This gives $\dot{x} = 3 \times 6^2 - 12 \times 6 = 36$ ms^{-1}.

7 (a) The velocity at time t is given by:
$\dot{x} = 12t^2 - 6t = 6t(2t - 1)$

(b) $\dot{x} = 0$ when $t = 0$ or $t = 0.5$ so the particle changes direction when $t = 0.5$
$t = 0 \Rightarrow x = 7$
$t = 0.5 \Rightarrow x = 6.75$
The distance travelled is 0.25 m.

(c) P is at its starting point when
$4t^3 - 3t^2 = 0$
$\Rightarrow t^2(4t - 3) = 0$
$\Rightarrow t = 0$, or $t = 0.75$
P returns to its start point after 0.75 s

Probability and Statistics 1

1 (a) Care is needed because the letter P is repeated. The key is to deal with this systematically. Any selection must contain 0, 1 or 2 Ps.

No. of Ps in selection	No. of selections
0	1 ($^4C_4$)
1	4 ($^4C_3$)
2	6 ($^4C_2$)
Total	11

(b)

No. of Ps in arrangement	No. of arrangements
0	$1 \times 4! = 24$
1	$4 \times 4! = 96$
2	$6 \times \dfrac{4!}{2} = 72$
Total	192

2 (a) $\Sigma P(X = x) = 1$
$\Rightarrow \dfrac{1}{k} + \dfrac{4}{k} + \dfrac{9}{k} = 1$
$\Rightarrow \dfrac{14}{k} = 1$
$k = 14$.

(b) Mean $= E(X) = \Sigma x P(X = x)$
$= 1 \times \dfrac{1}{14} + 2 \times \dfrac{4}{14} + 3 \times \dfrac{9}{14} = \dfrac{36}{14}$
$= \dfrac{18}{7}$.

$\text{Var}(X) = E(X^2) - (E(X))^2$
$= \Sigma x^2 P(X = x) - \left(\dfrac{18}{7}\right)^2$
$= \dfrac{1}{14} + 4 \times \dfrac{4}{14} + 9 \times \dfrac{9}{14} - \left(\dfrac{18}{7}\right)^2$
$= \dfrac{19}{49}$.

Practice examination answers

Probability and Statistics 1 (continued)

3 $X \sim B(20, 0.4)$ so $n = 20$ and $p = 0.4$

(a) Mean of $X = np = 20 \times 0.4 = 8$

Variance of $X = np(1-p) = 20 \times 0.4 \times 0.6$
$= 4.8$

(b) $P(2 \leq X \leq 4) = P(X=2) + P(X=3) + P(X=4)$.

$= \binom{20}{2}(0.4)^2(0.6)^{18} + \binom{20}{3}(0.4)^3(0.6)^{17}$

$+ \binom{20}{4}(0.4)^4(0.6)^{16}$

$= 0.00309 + 0.01235 + 0.03499 = 0.05043$

$= 0.050$ to 3 d.p.

Alternatively, use cumulative binomial tables

$P(2 \leq X \leq 4) = P(X \leq 4) - P(X \leq 1)$

$= 0.0510 - 0.0005 = 0.0505$

(c) $P(X \geq 5) = 1 - P(X \leq 4)$.

$= 1 - (P(X=0) + P(X=1) + P(2 \leq X \leq 4))$

$= 1 - (0.6^{20} + 20 \times 0.4 \times 0.6^{19} + 0.05043)$

$= 0.949$ to 3 d.p.

Alternatively, use cumulative binomial tables

$P(X \geq 5) = 1 - P(X \leq 4) = 1 - 0.0510 = 0.949$

4 (a) $P(X = 4) = p(1-p)^3 = 0.3(1-0.3)^3 = 0.1029$

(b) $P(X > 2) = (1-p)^2 = (1-0.3)^2 = 0.49$

(c) $E(X) = \dfrac{1}{p} = \dfrac{1}{0.3} = 3\dfrac{1}{3}$

5 (a) (i) $P(RRR) = \dfrac{26}{52} \times \dfrac{25}{51} \times \dfrac{24}{50} = \dfrac{2}{17}$

(ii) $P(\text{same colour}) = P(RRR) + P(BBB)$

$= \dfrac{2}{17} + \dfrac{2}{17} = \dfrac{4}{17}$.

(iii) $P(\text{at least one red}) = 1 - P(BBB)$

$= 1 - \dfrac{2}{17} = \dfrac{15}{17}$.

(b) Using A to represent getting an ace and D to represent getting a diamond:

$P(A|D') = \dfrac{P(A \cap D')}{P(D')} = \dfrac{\frac{3}{52}}{\frac{3}{4}} = \dfrac{3}{52} \times \dfrac{4}{3}$

$= \dfrac{1}{13}$.

6 (a)

Teacher A	Teacher B	Rankings A	B	d	d^2
39	46	5	5	0	0
72	80	1	1	0	0
51	53	3	4	−1	1
49	54	4	3	1	1
68	73	2	2	0	0
				$\Sigma d^2 = 2$	

$r_s = 1 - \dfrac{6 \times 2}{5^3 - 5} = 0.9$

(b) No. A high correlation produced in this way means that the markers agree closely in terms of the *relative* merit of the pieces of work but not necessarily in terms of the actual marks. For example, one marker may consistently award a higher mark than the other.

7 (a) Taking X to represent the number of packets in a box that contain a lucky pendant:

(i) $P(X \geq 6) = 1 - P(X \leq 5)$

Using tables of cumulative binomial probabilities with $n = 20$ and $p = 0.25$

$P(X \geq 6) = 1 - 0.6172 = 0.3828$

(ii) $P(X = 6) = P(X \leq 6) - P(X \leq 5)$

$= 0.7858 - 0.6172$

$= 0.1686$

(b) Using Y to represent the number of boxes that have at least six packets containing a lucky pendant:

$P(Y = 2) = \binom{5}{2}(0.3828)^2(1 - 0.3828)^3$

$= 0.34452 \ldots$

$= 0.345$ to 3 d.p.

Practice examination answers

Probability and Statistics 1 (continued)

8 (a) $r = \dfrac{S_{xy}}{\sqrt{S_{xx}S_{yy}}}$

$= \dfrac{28785 - \dfrac{567 \times 484}{10}}{\sqrt{\left(34216 - \dfrac{576^2}{10}\right)\left(24450 - \dfrac{484^2}{10}\right)}}$

$= 0.879$ to 3 d.p.

(b) $d = \dfrac{S_{xy}}{S_{yy}} = \dfrac{28785 - \dfrac{576 \times 484}{10}}{24450 - \dfrac{484^2}{10}}$

$= 0.885$ to 3 d.p.

$c = \dfrac{576}{10} - 0.8850 \times \dfrac{484}{10} = 14.766$

The equation is $x = 14.8 + 0.885y$.

(c) (i) $14.8 + 0.885 \times 42 = 51.97$

This gives an estimate of 52 marks.

(ii) A physics score of 80 is outside the interval of data for the explanatory variable so the result would be very unreliable.

Decision Mathematics 1

1 (a) Total memory to transfer = 4520 Kb = 4.52 Mb

$\dfrac{4.52}{1.4} = 3.228\ldots$

Three disks do not have sufficient capacity to hold all of the files. A minimum of four disks is required.

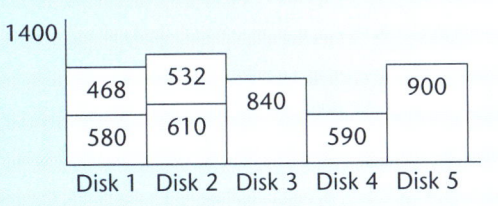

(b) [Bar chart showing: Disk 1: 468, 580; Disk 2: 532, 610; Disk 3: 840; Disk 4: 590; Disk 5: 900; with 1400 marked as capacity]

(c) Writing the memory sizes in order gives 900, 840, 610, 590, 580, 532, 468.

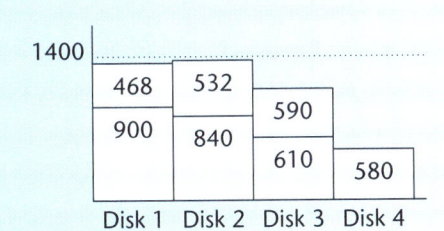

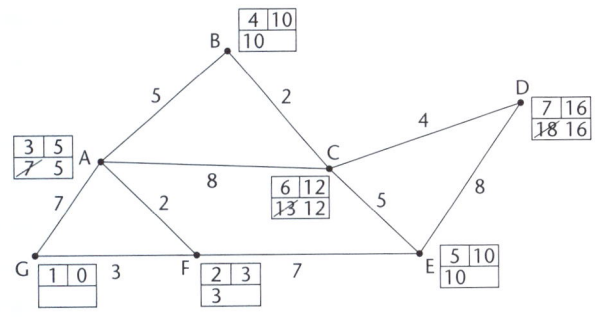

[Bar chart showing: Disk 1: 468, 900; Disk 2: 532, 840; Disk 3: 590, 610; Disk 4: 580; with 1400 marked as capacity]

2 [Network diagram with nodes labelled with working values]

(a) The shortest distance from G to D is 16.

(b) The route is G, F, A, B, C, D.

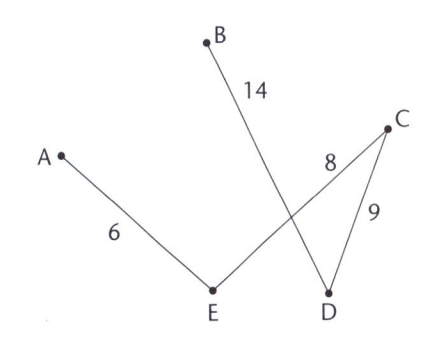

3 [Network diagram showing nodes A, B, C, D, E with edges AE=6, EC, CD=9, AB, BD=14, AC=8]

(a) The order in which the edges were connected is: AE, EC, CD, AB (not AD or AC to avoid cycles).

The length of the minimum spanning tree is 37.

(b) Using the short-cut from B to A gives an upper bound for the TSP as $37 + 12 = 49$.

Practice examination answers

Decision Mathematics 1 (continued)

4 (a)

	Vertices				
Graph (i)	A	B	C	D	E
Order	4	2	4	3	3

	Vertices					
Graph (ii)	A	B	C	D	E	F
Order	4	2	4	4	4	2

	Vertices			
Graph (iii)	A	C	D	E
Order	3	3	3	3

(b) Graph (i) is Semi-Eulerian (2 odd vertices).
Graph (ii) is Eulerian (all vertices are even).
Graph (iii) is neither (more than two odd vertices).

5 (a) The objective function is $4a + 9b + 12c$.

(b) The formulation of the task as a linear programming problem is:
Maximise: $P = 4a + 9b + 12c$
Subject to: $6a + 15b + 20c \leq 100$
$a + b + c \leq 10$
and $a \geq 0, b \geq 0, c \geq 0$.

(c) In the solution a, b and c must be integers.

6 (a) The initial tableau is:

P	x	y	z	s	t	val
1	−1	−2	−3	0	0	0
0	2	6	1	1	0	10
0	1	4	5	0	1	16

(b) The first iteration produces:

P	x	y	z	s	t	val
1	$-\frac{2}{5}$	$\frac{2}{5}$	0	0	$\frac{3}{5}$	$9\frac{3}{5}$
0	$\frac{9}{5}$	$\frac{26}{5}$	0	1	$-\frac{1}{5}$	$6\frac{4}{5}$
0	$\frac{1}{5}$	$\frac{4}{5}$	1	0	$\frac{1}{5}$	$3\frac{1}{5}$

The second iteration produces:

P	x	y	z	s	t	val
1	0	$\frac{14}{9}$	0	$\frac{2}{9}$	$\frac{5}{9}$	$11\frac{1}{9}$
0	1	$\frac{26}{9}$	0	$\frac{5}{9}$	$-\frac{1}{9}$	$3\frac{7}{9}$
0	0	$\frac{2}{9}$	1	$-\frac{1}{9}$	$\frac{2}{9}$	$2\frac{4}{9}$

The maximum value of P is $11\frac{1}{9}$.
This occurs when $x = 3\frac{7}{9}$, $y = 0$, $z = 2\frac{4}{9}$.

(c) This represents the optimum solution because there are no negative values in the objective row of the tableau.

Notes

Notes

Notes

Notes

Index

acceleration 58–60, 66–67
addition rule 78
algebra 13–21, 36–39
algebraic division 37–38
algorithms 91–98
arc 46, 94
area of a triangle 45
area under a curve 50–51
arithmetic progression (series) 41
arrangements 80
asymptote 18
average 75

bin packing 93
binomial expansion 42–43
bounds 97
box and whisker plot 76
bubble sort 92

calculus 46, 49
circle 25
circle properties 26
coefficient of friction 64
collision 67
combinations 80
common difference 40
common ratio 40
complete graph 94
completing the square 15
conditional probability 79
connected graph 94
connected particles 66
conservation of momentum 67
constant acceleration 58–59
constraints 99
coordinate geometry 23–26
correlation 84
correlation coefficient 84
cosine 44
cosine rule 45
cubic curve 17
cumulative frequency 77
curve sketching 29
cycle 94

data 75–76
deceleration 60
decreasing function 29
definite integration 50
degree 94
degree of a polynomial 17
dependent variable 85

derivative 28–30
differentiation 28–30
digraph 94
Dijkstra's algorithm 96
directed edge 94
discrete random variable 81
discriminant 16
dispersion 76
displacement 58–60
distance 58–59
distance between two points 25
dynamics 66–68

equation of circle 25
equilibrium 64
Eulerian cycle 94
Eulerian graph 94, 98
even vertex 94
event 78
expectation (expected value) 81
exponential functions 36
extrapolation 85

factor theorem 38–39
factorising 14–15, 38–39
feasible region 99
first-fit algorithm 93
flow diagram 91
forces 58, 62–67
frequency density 77
frequency distribution 77
frequency polygon 77
friction 64
function 14

geometric distribution 83
geometric series 41
gradient 23, 28, 59
gradient function 28
gradient-intercept form 23
graphical terms 94
gravity 59
grouped data 75

histogram 77

identity 39, 47
image 14
impact 69
inclined plane 62
increasing function 29
independent events 79

independent variable 85
indices 13
inequalities 20
integral 49–52
integration 49–52
integration constant 49
intercept 23
interquartile range 76
iteration 101

kinematics 58–61
Kruskal's algorithm 95

least squares regression line 85
light inextensible string 66
limiting equilibrium 64
linear programming 99–101
log laws 37
logarithm 36
lower quartile 76

magnitude 58
map 14
mass 58, 66–67
maximum of a function 29–30
mean 75–76, 81
median 75–76
mid-point 25
minimum of a function 29–30
minimum connector 94
minimum spanning tree 94, 97
modal class 77
mode 75
momentum 58, 67
multiplication rule 79
mutually exclusive events 78

nearest neighbour algorithm 97
networks 94, 98
Newton's laws of motion 66–67
nodes 94
normal reaction 62
normal to a curve 29
numerical integration 51–52

objective function 99
odd vertex 94
optimum tableau 101
order of an algorithm 93
outcome 78
outlier 75, 85

Index

parabola 16
parallel lines 23
parameter 82
Pascal's triangle 42–43
pass 92
path 94
periodic function 44
permutations 80
perpendicular lines 24
pivotal column and row 100
planar graph 94
point of inflexion 29
polynomials 17
Prim's algorithm 95
probability 78–80
probability distribution 81
product-moment correlation
    coefficient 84

quadratic formula 15
quadratic functions and equations
    14–17
quadratic graphs 16
quartiles 76
quotient 37

radians 46
random variable 81–83
range 76
rationalising the denominator 13
reaction 62, 66
recurrence relation 40
recursive definition 40
reflection 37
reflective symmetry 18
regression 85
remainder 37
remainder theorem 38

residuals 85
resolving forces 62
resultant force 63
roots of equations 16
rotational symmetry 18, 44
route inspection problem 98

sample space 78
scalar quantity 58
scatter diagram 84
second derivative 29–30
semi-Eulerian 94, 98
sequences 40
series 40–41
sets 91
shuttle sort 92
Simplex algorithm 100
simultaneous equations 19
sine 44
sine rule 45
skew 76
slack variables 100
spanning tree 94
Spearman's rank correlation
    coefficient 84
speed 58–59
standard deviation 76, 81
stationary points 29–30
statistical diagrams 76–77
stem and leaf diagram 76
straight line equation 23–24
straight line motion 58–59
subgraph 94
substitution 19
sum of a series 40–41
surds 13
symmetry 18, 44

tableau 100
tangent function 28–29, 44–47
tangent to a curve 26, 28
tension 62, 66
transformations 19, 45
trapezium rule 51–52
travelling salesperson problem 97
traversable graph 98
tree 94
tree diagram 79
triangle area 45
trigonometry 44–47

upper quartile 76

valency 94
variable acceleration 60
variance 76, 81
vector diagrams 62
vectors 58
velocity 58–60
Venn diagrams 78
vertex 17, 94

weight 62

Letts

Revise
A2

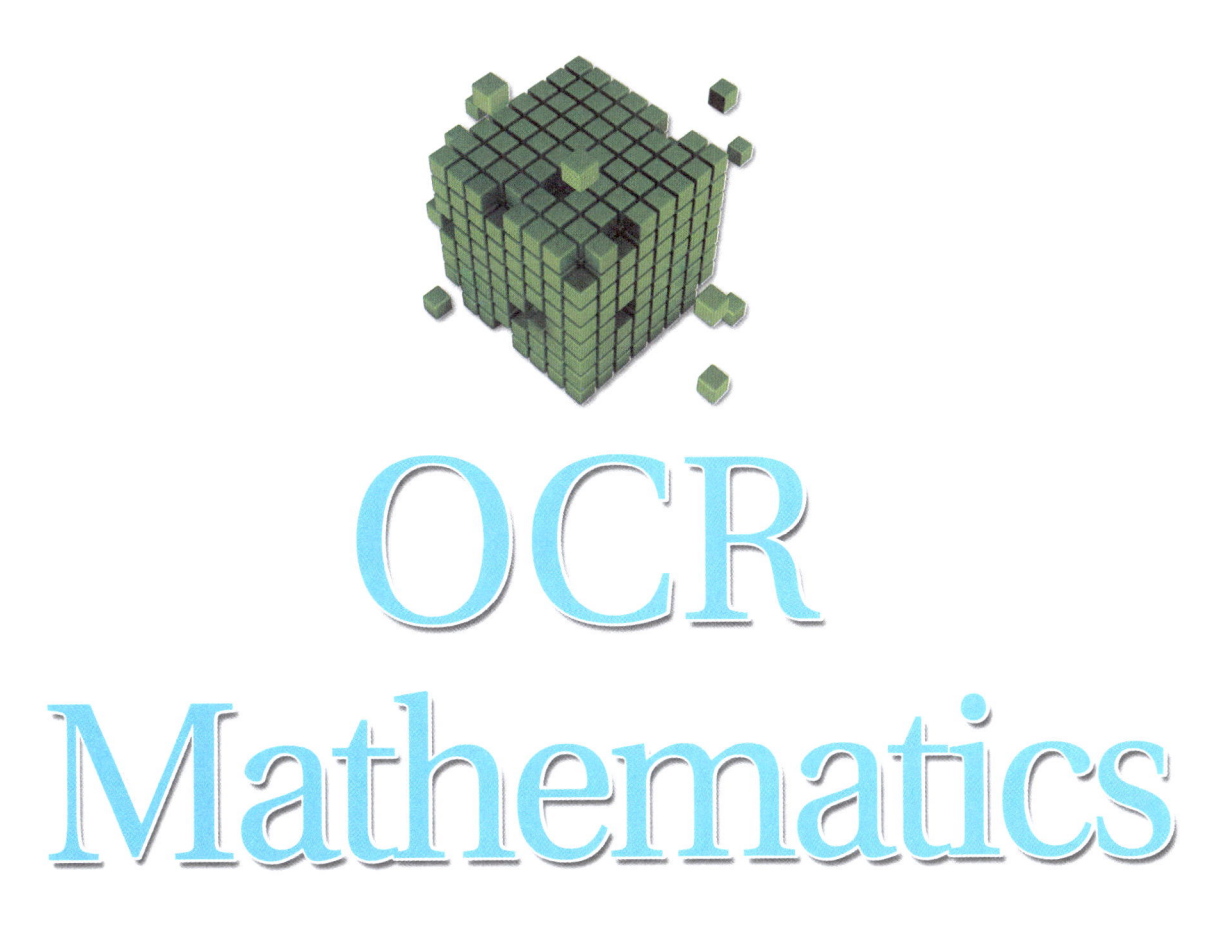

OCR
Mathematics

Peter Sherran & Janet Crawshaw

Contents

Specification information	4
Your AS/A2 Level Mathematics course	5
Exam technique	8
What grade do you want?	11
Four steps to successful revision	13

Chapter 1 Core 3 and Core 4 (Pure Mathematics)

1.1	Algebra and series	14
1.2	Functions including exponential and logarithmic	18
1.3	Coordinate geometry	25
1.4	Trigonometry	27
1.5	Differentiation	31
1.6	Integration	39
1.7	Numerical methods	49
1.8	Vectors	53
	Sample questions and model answers	58
	Practice examination questions	62

Chapter 2 Probability and Statistics 2

2.1	Continuous random variables	64
2.2	The Poisson distribution and approximations	67
2.3	Estimation and sampling	71
2.4	Hypothesis tests	74
	Sample questions and model answers	79
	Practice examination questions	82

Chapter 3 Mechanics 2

3.1	Projectiles	84
3.2	Equilibrium of a rigid body	87
3.3	Centre of mass	90
3.4	Collisions and impulse	93
3.5	Uniform circular motion	97
3.6	Work, energy and power	99
	Sample questions and model answers	101
	Practice examination questions	105

Chapter 4 Decision Mathematics 2

4.1	Game theory	109
4.2	Networks	113
4.3	Critical path analysis	117
4.4	Matching and allocation	121
4.5	Dynamic programming	125
	Sample questions and model answers	128
	Practice examination questions	130

Practice examination answers	132
Index	143

Specification information

OCR Mathematics

MODULE	SPECIFICATION TOPIC	CHAPTER REFERENCE	STUDIED IN CLASS	REVISED	PRACTICE QUESTIONS
Core 3 (C3)	Algebra and functions	1.2, 1.6			
	Trigonometry	1.4			
	Differentiation and integration	1.5, 1.6			
	Numerical methods	1.7			
Core 4 (C4)	Algebra and graphs	1.1, 1.3			
	Differentiation and integration	1.5, 1.6			
	First order differential equations	1.6			
	Vectors	1.8			
Probability and Statistics 2 (S2)	Continuous random variables	2.1			
	The normal distribution	2.1, 2.2			
	The Poisson distribution	2.2			
	Sampling and hypothesis tests	2.3, 2.4			
Mechanics 2 (M2)	Centre of mass	3.3			
	Equilibrium of a rigid body	3.2			
	Motion of a projectile	3.1			
	Uniform motion in a circle	3.5			
	Coefficient of restitution; impulse	3.4			
	Energy, work and power	3.6			
Decision 2 (D2)	Game theory	4.1			
	Flows in a network	4.2			
	Matching and allocation problems	4.4			
	Critical path analysis	4.3			
	Dynamic programming	4.5			

Examination analysis

The Advanced Level Mathematics GCE consists of AS (50%) and A2 (50%). AS consists of C1 + C2 + one of M1, S1, D1. A2 consists of C3 + C4 + one of M1, S1, D1, M2, S2, D2 (following any dependency rules).

C3	A2	Scientific/graphics calculator	1 hr 30 min exam	$16\frac{2}{3}$% of A Level
C4	A2	Scientific/graphics calculator	1 hr 30 min exam	$16\frac{2}{3}$% of A Level
M1/M2	AS/A2	Scientific/graphics calculator	1 hr 30 min exam	$16\frac{2}{3}$% of A Level
S1/S2	AS/A2	Scientific/graphics calculator	1 hr 30 min exam	$16\frac{2}{3}$% of A Level
D1/D2	AS/A2	Scientific/graphics calculator	1 hr 30 min exam	$16\frac{2}{3}$% of A Level

Your AS/A2 Level Mathematics course

AS and A2

The OCR Mathematics A Level course is in two parts, with three separate modules in each part. Students first study the AS (Advanced Subsidiary) course. Some will then go on to study the second part of the A Level course, called A2. Advanced Subsidiary is assessed at the standard expected halfway through an A Level course: i.e., between GCSE and Advanced GCE. This means that the AS and A2 courses are designed so that difficulty steadily increases:

- AS Mathematics builds from GCSE Mathematics
- A2 Mathematics builds from AS Mathematics.

How will you be tested?

Assessment units

For AS Mathematics, you will be tested by three assessment units. For the full A Level in Mathematics, you will take a further three units. AS Mathematics forms 50% of the assessment weighting and A2 Mathematics forms the other 50% for the full A Level.

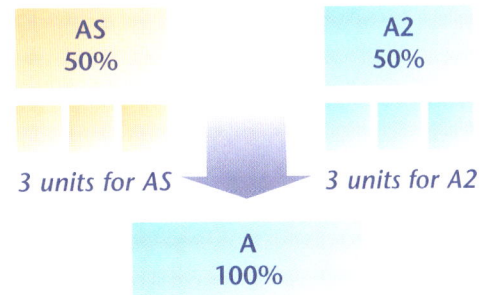

```
     AS              A2
    50%             50%

3 units for AS    3 units for A2

          A
         100%
```

Most units can be taken in either January or June, but some A2 units must be taken at the end of the course. There is a lot of flexibility about when exams can be taken and the diagram below shows just some of the ways that the assessment units may be taken for AS and A Level Mathematics.

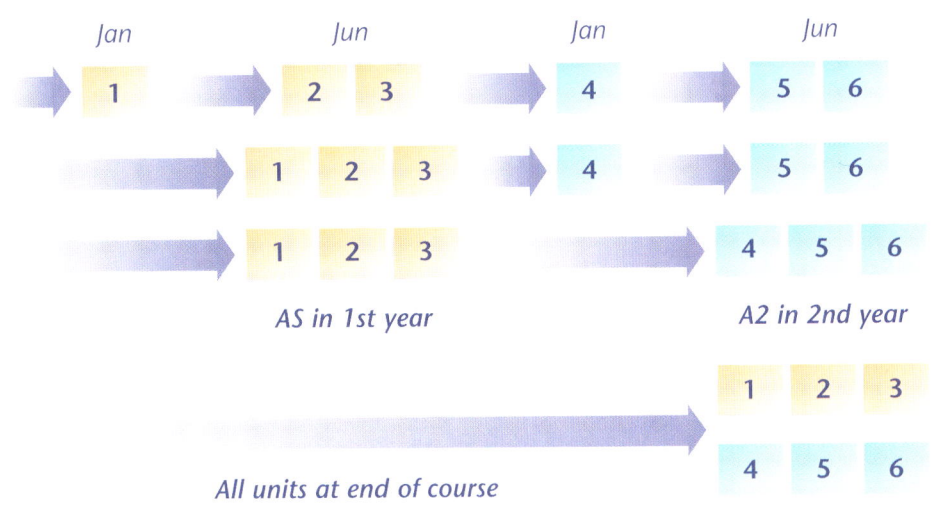

```
    Jan         Jun         Jan         Jun
     1         2  3          4         5  6
              1  2  3         4         5  6
                 1  2  3                4  5  6

         AS in 1st year              A2 in 2nd year

                                     1  2  3
                                     4  5  6

         All units at end of course
```

5

If you are disappointed with a module result, you can resit the module. There is no restriction on the number of times a module may be attempted. The best available results for each module will count towards the final grade.

A2 progression

After having studied AS Mathematics, to continue studying Mathematics to A Level you will need to take three further units of Mathematics, at least two of which are at A2 standard.

Synoptic assessment

Synoptic assessment in Mathematics addresses candidates' understanding of the connections between different elements of the subject. It involves the explicit drawing together of knowledge, understanding and skills learned in different parts of the A Level course, through using and applying methods developed earlier to solving problems. Making and understanding connections in this way is intrinsic to learning mathematics.

For A2 modules you will be asked to demonstrate knowledge of modules studied at AS Level. The contents of C1 and C2 are assumed for C3 and the contents of C1, C2 and C3 are assumed for C4. Similarly in the application modules, a knowledge of S1 is assumed for S2, a knowledge of M1 is assumed for M2 and a knowledge of D1 is assumed for D2.

Key skills

It is important that you develop your key skills throughout your AS and A2 courses. These are important skills that you need whatever you do beyond AS and A Levels. To gain the key skills qualification, which is equivalent to an AS Level, you will need to collect evidence together in a 'portfolio' to show that you have attained Level 3 in Communication, Application of number and Information technology. You will also need to take a formal test in each key skill. You will have many opportunities during AS and A2 Mathematics to develop your key skills.

It is a worthwhile qualification, as it demonstrates your ability to put your ideas across to other people, collect data and use up-to-date technology in your work.

Your AS/A2 Level Mathematics course

What skills will I need?

For A Level Mathematics (AS and A2), you will be tested by assessment objectives: these are the skills and abilities that you should have acquired by studying the course. The assessment objectives for A Level Mathematics are shown below.

Candidates should be able to:

- recall, select and use their knowledge of mathematical facts, concepts and techniques in a variety of contexts

- construct rigorous mathematical arguments and proofs through use of precise statements, logical deduction and inference and by the manipulation of mathematical expressions, including the construction of extended arguments for handling substantial problems presented in unstructured form

- recall, select and use their knowledge of standard mathematical models to represent situations in the real world; recognise and understand given representations involving standard models; present and interpret results from such models in terms of the original situation, including discussion of the assumptions made and refinement of such models

- comprehend translations of common realistic contexts into mathematics; use the results of calculations to make predictions, or comment on the context; and, where appropriate, read critically and comprehend longer mathematical arguments or examples of applications

- use contemporary calculator technology and other permitted resources (such as formulae booklets or statistical tables*) accurately and efficiently; understand when not to use such technology, and its limitations; give answers to appropriate accuracy.

* You can find a copy of the formulae booklet and statistical tables in the A Level mathematics section of the OCR website, www.ocr.org.uk

Exam technique

What are the examiners looking for?

Examiners use certain words in their instructions to let you know what they are expecting in your answer. Make sure that you know what they mean so that you can give the right response.

Write down, state

You can write your answer without having to show how it was obtained. There is nothing to prevent you doing some working if it helps you, but if you are doing a lot then you might have missed the point.

Calculate, find, determine, show, solve

Make sure that you show enough working to justify the final answer or conclusion. Marks will be available for showing a correct method.

Deduce, hence

This means that you are expected to use the given result to establish something new. You must show all of the steps in your working.

Draw

This is used to tell you to plot an accurate graph using graph paper. Take note of any instructions about the scale that must be used. You may need to read values from your graph.

Sketch

If the instruction is to sketch a graph then you don't need to plot the points but you will be expected to show its general shape and its relationship with the axes. Indicate the positions of any turning points and take particular care with any asymptotes.

Find the exact value

This instruction is usually given when the final answer involves an irrational value such as a logarithm, e, π or a surd. You will need to demonstrate that you can manipulate these quantities so don't just key everything into your calculator or you will lose marks.

If a question requires the final answer to be given to a specific level of accuracy then make sure that you do this or you might needlessly lose marks.

Exam technique

Some dos and don'ts

Dos

Do read the question

- Make sure that you are clear about what you are expected to do. Look for some structure in the question that may help you take the right approach.
- Read the question *again* after you have answered it as a quick check that your answer is in the expected form.

Do use diagrams

- In some questions, particularly in mechanics, a clearly labelled diagram is essential. Use a diagram whenever it may help you understand or represent the problem that you are trying to solve.

Do take care with notation

- Write clearly and use the notation accurately. Use brackets when they are required.
- Even if your final answer is wrong, you may earn some marks for a correct expression in your working.

Do learn relevant formulae

- Each module specification contains a list of formulae that *will not be given* in the examination. Make sure that you learn these formulae and practise using them.
- Some formulae *will be given* in the Examination Formulae Booklet. Make sure you know what they are and where to find them. This will help you refer to them quickly in the examination so that you don't waste time.

Do avoid silly answers

- Check that your final answer is sensible within the context of the question.

Do make good use of time

- Choose the order in which you answer the questions carefully. Do the ones that are easiest for you first.
- Set yourself a time limit for a question depending on the number of marks available.
- Be prepared to leave a difficult part of a question and return to it later if there is time.
- Towards the end of the exam make sure that you pick up all of the easy marks in any questions that you haven't got time to answer fully.

Don'ts

Don't work with rounded values

- There may be several stages in a solution that produce numerical values. Rounding errors from earlier stages may distort your final answer. One way to avoid this is to make use of your calculator memories to store values that you will need again.

Exam technique

Don't cross out work that may be partly correct

- It's tempting to cross out something that hasn't worked out as it should. Avoid this unless you have time to replace it with something better.

Don't write out the question

- This wastes time. The marks are for your solution!

What grade do you want?

Everyone should be able to improve their grades but you will only manage this with a lot of hard work and determination. The details given below describe a level of performance typical of candidates achieving grades A, C or E. You should find it useful to read and compare the expectations for the different levels and to give some thought to the areas where you need to improve most.

Grade A candidates

- Recall or recognise almost all the mathematical facts, concepts and techniques that are needed, and select appropriate ones to use in a variety of contexts.
- Manipulate mathematical expressions and use graphs, sketches and diagrams, all with high accuracy and skill.
- Use mathematical language correctly and proceed logically and rigorously through extended arguments or proofs.
- When confronted with unstructured problems they can often devise and implement an effective solution strategy.
- If errors are made in their calculations or logic, these are sometimes noticed and corrected.
- Recall or recognise almost all the standard models that are needed, and select appropriate ones to represent a wide variety of situations in the real world.
- Correctly refer results from calculations using the model to the original situation; they give sensible interpretations of their results in the context of the original realistic situation.
- Make intelligent comments on the modelling assumptions and possible refinements to the model.
- Comprehend or understand the meaning of almost all translations into mathematics of common realistic contexts.
- Correctly refer the results of calculations back to given context and usually make sensible comments or predictions.
- Can distil the essential mathematical information from extended pieces of prose having mathematical content.
- Comment meaningfully on the mathematical information.
- Make appropriate and efficient use of contemporary calculator technology and other permitted resources, and are aware of any limitations to their use.
- Present results to an appropriate degree of accuracy.

Grade C candidates

- Recall or recognise most of the mathematical facts, concepts and techniques that are needed, and usually select appropriate ones to use in a variety of contexts.
- Manipulate mathematical expressions and use graphs, sketches and diagrams, all with a reasonable level of accuracy and skill.
- Use mathematical language with some skill and sometimes proceed logically through extended arguments or proofs.
- When confronted with unstructured problems they sometimes devise and implement an effective and efficient solution strategy.
- Occasionally notice and correct errors in their calculations.
- Recall or recognise most of the standard models that are needed and usually select appropriate ones to represent a variety of situations in the real world.

What grade do you want?

- Often correctly refer results from calculations using the model to the original situation; they sometimes give sensible interpretations of their results in context of the original realistic situation.
- Sometimes make intelligent comments on the modelling assumptions and possible refinements to the model.
- Comprehend or understand the meaning of most translations into mathematics of common realistic contexts.
- Often correctly refer the results of calculations back to the given context and sometimes make sensible comments or predictions.
- Distil much of the essential mathematical information from extended pieces of prose having mathematical content.
- Give some useful comments on this mathematical information.
- Usually make appropriate and effective use of contemporary calculator technology and other permitted resources, and are sometimes aware of any limitations to their use.
- Usually present results to an appropriate degree of accuracy.

Grade E candidates

- Recall or recognise some of the mathematical facts, concepts and techniques that are needed, and sometimes select appropriate ones to use in some contexts.
- Manipulate mathematical expressions and use graphs, sketches and diagrams, all with some accuracy and skill.
- Sometimes use mathematical language correctly and occasionally proceed logically through extended arguments or proofs.
- Recall or recognise some of the standard models that are needed and sometimes select appropriate ones to represent a variety of situations in the real world.
- Sometimes correctly refer results from calculations using the model to the original situation; they try to interpret their results in the context of the original realistic situation.
- Sometimes comprehend or understand the meaning of translations into mathematics of common realistic contexts.
- Sometimes correctly refer the results of calculations back to the given context and attempt to give comments or predictions.
- Distil some of the essential mathematical information from extended pieces of prose having mathematical content; they attempt to comment on this mathematical information.
- Candidates often make appropriate and efficient use of contemporary calculator technology and other permitted resources.
- Often present results to an appropriate degree of accuracy.

The table below shows how your uniform standardised mark is translated.

UMS	80%	70%	60%	50%	40%
grade	A	B	C	D	E

To achieve an A* grade in Mathematics, you need to achieve a grade A overall (an average of 80 or more on uniform mark scale) for the whole A Level qualification and an average of 90 or more on the uniform mark scale in Core 3 and Core 4. It is awarded for A Level qualification only and not for the AS qualification or individual units.

Four steps to successful revision

Step 1: Understand

- Study the topic to be learned slowly. Make sure you understand the logic or important concepts.
- Mark up the text if necessary – underline, highlight and make notes.
- Re-read each paragraph slowly.

GO TO STEP 2

Step 2: Summarise

- Now make your own revision note summary:
  What is the main idea, theme or concept to be learned?
  What are the main points? How does the logic develop?
  Ask questions: Why? How? What next?
- Use bullet points, mind maps, patterned notes.
- Link ideas with mnemonics, mind maps, crazy stories.
- Note the title and date of the revision notes
  (e.g. Mathematics: Differentiation, 3rd March).
- Organise your notes carefully and keep them in a file.

This is now in **short-term memory**. You will forget 80% of it if you do not go to Step 3.
GO TO STEP 3, but first take a 10 minute break.

Step 3: Memorise

- Take 25 minute learning 'bites' with 5 minute breaks.
- After each 5 minute break test yourself:
  Cover the original revision note summary
  Write down the main points
  Speak out loud (record yourself)
  Tell someone else
  Repeat many times.

The material is well on its way to **long-term memory**.
You will forget 40% if you do not do step 4. **GO TO STEP 4**

Step 4: Track/Review

- Create a Revision Diary (one A4 page per day).
- Make a revision plan for the topic, e.g. 1 day later, 1 week later, 1 month later.
- Record your revision in your Revision Diary, e.g.
  Mathematics: Differentiation, 3rd March 25 minutes
  Mathematics: Differentiation, 5th March 15 minutes
  Mathematics: Differentiation, 3rd April 15 minutes
  ... and then at monthly intervals.

Chapter 1
Core 3 and Core 4 (Pure Mathematics)

The following topics are covered in this chapter:

- Algebra and series
- Functions including exponential and logarithmic
- Coordinate geometry
- Trigonometry
- Differentiation
- Integration
- Numerical methods
- Vectors

1.1 Algebra and series

LEARNING SUMMARY

After studying this section you should be able to:
- simplify rational expressions
- express rational expressions in partial fractions
- use the binomial expansion $(1+x)^n$ for any rational n

Rational expressions

Core 4

A **rational expression** is of the form $\dfrac{f(x)}{g(x)}$, where $f(x)$ and $g(x)$ are polynomials in x.

To simplify a rational expression, factorise both $f(x)$ and $g(x)$ as far as possible, then cancel any factors that appear in both the numerator and denominator.

Examples

(a) $\dfrac{2x^2 + 6x}{2x^2 + 7x + 3} = \dfrac{2x(x+3)}{(2x+1)(x+3)}$

$= \dfrac{2x}{2x+1}$ — This cannot be cancelled any further.

> Factorise $x^2 - 4$ using difference between two squares.

(b) $\dfrac{3x^2 - 12}{2 - x} = \dfrac{3(x^2 - 4)}{2 - x}$

> $x - 2 = -(2 - x)$.

$= \dfrac{3(x-2)(x+2)}{(2-x)}$

$= -3(x+2)$

When adding and subtracting algebra fractions, always use the lowest common denominator, for **example**,

> First factorise the denominators.
> The lowest common multiple of the denominators is $(x+3)(x-4)(x-3)$.

$\dfrac{x-2}{x^2 - x - 12} + \dfrac{4}{x^2 - 9} = \dfrac{x-2}{(x+3)(x-4)} + \dfrac{4}{(x+3)(x-3)}$

$= \dfrac{(x-3)(x-2) + 4(x-4)}{(x+3)(x-4)(x-3)}$

$= \dfrac{x^2 - 5x + 6 + 4x - 16}{(x+3)(x-4)(x-3)}$

$= \dfrac{x^2 - x - 10}{(x+3)(x-4)(x-3)}$

Key points from AS
- Remainder theorem page 38
- Algebraic division page 37

14

Core 3 and Core 4 (Pure Mathematics)

Partial fractions

Core 4

It is sometimes possible to decompose a rational function into **partial fractions**. This format is often useful when differentiating, integrating or expanding functions as a series. Look out for the following types of partial fractions.

Denominator contains linear factors only

$$\frac{x+7}{(x-2)(x+1)} \equiv \frac{A}{x-2} + \frac{B}{x+1} = \frac{A(x+1) + B(x-2)}{(x-2)(x+1)}$$

> The symbol $\equiv$ indicates that this is an identity. It is true for all values of x.

$$\Rightarrow x + 7 \equiv A(x+1) + B(x-2)$$

Let $x = -1$, then $6 = -3B \Rightarrow B = -2$
Let $x = 2$, then $9 = 3A \Rightarrow A = 3$

> These substitutions are chosen so that each factor in turn becomes zero.

$$\therefore \frac{x+7}{(x-2)(x+1)} \equiv \frac{3}{x-2} - \frac{2}{x+1}$$

Denominator contains repeated linear factors

> Do not use $(x+3)(x+1)(x+1)^2$ as the common denominator as it is not the lowest common multiple.

$$\frac{6x^2 + 15x + 7}{(x+3)(x+1)^2} \equiv \frac{A}{x+3} + \frac{B}{x+1} + \frac{C}{(x+1)^2}$$

$$\equiv \frac{A(x+1)^2 + B(x+3)(x+1) + C(x+3)}{(x+3)(x+1)^2}$$

$$\Rightarrow 6x^2 + 15x + 7 \equiv A(x+1)^2 + B(x+3)(x+1) + C(x+3)$$

> Use a mixture of substitution and equating coefficients to find A, B and C.

Let $x = -1$, then $-2 = 2C \Rightarrow C = -1$
Let $x = -3$, then $16 = 4A \Rightarrow A = 4$
Equate x^2 terms: $6 = A + B$ and since $A = 4$, $B = 2$.

$$\therefore \frac{6x^2 + 15x + 7}{(x+3)(x+1)^2} \equiv \frac{4}{x+3} + \frac{2}{x+1} - \frac{1}{(x+1)^2}$$

> The degree of a polynomial is its highest power of x, for example $x^3 + 2x^2 - 3x + 4$ has degree 3.

Only **proper fractions** (when the numerator is of lower degree than the denominator) can be written in partial fraction form. For **improper fractions**, divide the denominator into the numerator first. This can be done by long division or by using identities.

Binomial expansion for rational n

Core 4

> **KEY POINT**
>
> When n is rational, $(1 + x)^n$ can be written as a series, using the **binomial expansion**
>
> $$(1 + x)^n = 1 + nx + \frac{n(n-1)}{2!}x^2 + \frac{n(n-1)(n-2)}{3!}x^3 + \ldots$$

Key points from AS

- Binomial expansion when n is a positive integer
  Revise AS page 42

If n is a positive integer, the series terminates at the term in x^n.

For all other values of n, the series is infinite and converges provided that $|x| < 1$.

Example

(a) Expand $\sqrt{1-4x}$ as an ascending series in x, as far as the term in x^3, giving the set of values of x for which the series is valid.

(b) By substituting $x = 0.01$ into the series expansion of $\sqrt{1-4x}$, find $\sqrt{96}$ correct to 4 decimal places.

> $1 - 4x = 1 + (-4x)$, so use $(1+x)^n$ with $n = \frac{1}{2}$ and substituting $(-4x)$ for x.
>
> You should also state the set of values of x for which the expansion is valid, even when it is not specifically requested in the question.

(a) $\sqrt{1-4x} = (1-4x)^{\frac{1}{2}}$

$$= 1 + \tfrac{1}{2}(-4x) + \frac{(\tfrac{1}{2})(-\tfrac{1}{2})}{2!}(-4x)^2 + \frac{(\tfrac{1}{2})(-\tfrac{1}{2})(-\tfrac{3}{2})}{3!}(-4x)^3 + \ldots$$

$$= 1 - 2x - 2x^2 - 4x^3 + \ldots$$

The series is valid provided that $|4x| < 1$, i.e. $|x| < \tfrac{1}{4}$.

(b) Substituting $x = 0.01$ into $\sqrt{1-4x}$ gives $\sqrt{0.96} = \sqrt{\dfrac{96}{100}} = \dfrac{\sqrt{96}}{10}$.

So $\sqrt{96} = 10(1 - 2(0.01) - 2(0.01)^2 - 4(0.01)^3 + \ldots)$
$= 10(1 - 0.02 - 0.0002 - 0.000004 +)$
$= 9.7980$ (4 d.p.)

The function to be expanded must be in the form $(1 + \ldots)^n$ or $(1 - \ldots)^n$.
You may have to do some careful algebraic manipulation to get it in this form.

Example

Give the first four terms in the series expansion of $\dfrac{1}{(2+x)}$.

$$\frac{1}{(2+x)} = \frac{1}{2\left(1+\frac{x}{2}\right)}$$

> Alternatively, $(2+x)^{-1} = 2^{-1}\left(1+\frac{x}{2}\right)^{-1}$.

$$= \frac{1}{2}\left(1+\frac{x}{2}\right)^{-1}$$

$$= \frac{1}{2}\left(1 + (-1)\left(\frac{x}{2}\right) + \frac{(-1)(-2)}{2!}\left(\frac{x}{2}\right)^2 + \frac{(-1)(-2)(-3)}{3!}\left(\frac{x}{2}\right)^3 + \ldots\right)$$

$$= \frac{1}{2}\left(1 - \frac{x}{2} + \frac{x^2}{4} - \frac{x^3}{8} + \ldots\right)$$

$$= \frac{1}{2} - \frac{x}{4} + \frac{x^2}{8} - \frac{x^3}{16} + \ldots$$

Restriction on x: $\left|\dfrac{x}{2}\right| < 1$, i.e. $|x| < 2$.

Core 3 and Core 4 (Pure Mathematics)

Progress check

Core 4

1 Simplify:

(a) $\dfrac{2x^2 + 3x}{2x^2 + x - 3}$ (b) $\dfrac{9x - x^3}{3x^2 + x^3}$

Core 4

2 Express as a single fraction in its simplest form:

(a) $\dfrac{3(x - 4)}{(x + 2)(x - 1)} + \dfrac{2(x + 1)}{x - 1}$

(b) $\dfrac{x^2 - 3x + 2}{x^2 + 3x - 4} - \dfrac{1}{(x + 4)^2}$

Core 4

3 $\dfrac{4x^2 + x + 1}{(x + 2)(x - 3)(x + 1)} \equiv \dfrac{A}{x + 2} + \dfrac{B}{x - 3} + \dfrac{C}{x + 1}$

Find the values of A, B and C.

Core 4

4 Express in partial fractions:

$\dfrac{3x^2 + 4x + 5}{(x - 1)^2(x + 2)}$

Core 4

5 Write as a series in ascending powers of x as far as the term in x^3, stating the values of x for which the expansion is valid:

(a) $\dfrac{1}{1 + x}$ (b) $(1 - 3x)^{-2}$ (c) $\sqrt{4 + x}$

---

1 (a) $\dfrac{x}{x-1}$ (b) $\dfrac{x}{3-x}$

2 (a) $\dfrac{2x^2 + 9x - 8}{(x + 2)(x - 1)}$ (b) $\dfrac{x^2 + 2x - 9}{(x + 4)^2}$

3 $A = 3,\ B = 2,\ C = -1$

4 $\dfrac{2}{x - 1} + \dfrac{4}{(x - 1)^2} + \dfrac{1}{x + 2}$

5 (a) $1 - x + x^2 - x^3 + \ldots$ for $|x| < 1$
   (b) $1 + 6x + 27x^2 + 108x^3 + \ldots$ for $|x| < \tfrac{1}{3}$
   (c) $2 + \tfrac{1}{4}x - \tfrac{1}{64}x^2 + \tfrac{1}{512}x^3 + \ldots$ for $|x| < 4$

Core 3 and Core 4 (Pure Mathematics)

1.2 Functions including exponential and logarithmic

LEARNING SUMMARY

After studying this section you should be able to:

- use function notation and understand domain and range
- understand composite functions and find the inverse of a one-one function
- sketch the graph of the inverse of a function from the graph of the function
- understand the properties of the modulus function
- use transformations to sketch graphs of related functions
- understand the definitions of e^x and $\ln x$

Functions

Core 3

A **function** may be thought of as a rule which takes each member x of a set and assigns, or **maps**, it to some value y known as its **image**.

x maps to y
y is the image of x.

$x \longrightarrow \boxed{\text{Function}} \longrightarrow y$

f(x) is read as 'f of x'.

A letter such as f, g or h is often used to stand for a function. The function which squares a number and adds on 5, for example, can be written as $f(x) = x^2 + 5$. The same notation may also be used to show how a function affects particular values. For this function, $f(4) = 4^2 + 5 = 21$, $f(-10) = (-10)^2 + 5 = 105$ and so on.

An alternative notation for the same function is $f: x \mapsto x^2 + 5$.

KEY POINT

The set of values on which the function acts is called the **domain** and the corresponding set of image values is called the **range**.

The function $f(x) = x + 2$ with domain {3, 6, 10} is shown below.

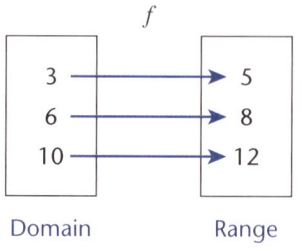

f

3 → 5
6 → 8
10 → 12

Domain   Range

The symbol ℝ is often used to stand for the complete set of numbers on the number line. This is the set of Real numbers.

The domain of a function may be an infinite set such as ℝ but, in some cases, particular values must be omitted for the function to be valid.

For example, the function $f(x) = \dfrac{x+2}{x-3}$ cannot have 3 in its domain since division by zero is undefined.

Under a function, every member of the domain must have only one image. But, it is possible for different values in the domain to have the same image.

There are different types of function.

KEY POINT

If for each element y in the range there is *a unique* value of x such that $f(x) = y$, then f is a **one–one** function. If for any element y in the range there is more than one value of x satisfying $f(x) = y$, then f is **many–one**.

Core 3 and Core 4 (Pure Mathematics)

Examples

The function $f(x) = 2x$ with domain $\mathbb{R}$ is one–one.

The function $f(x) = x^2$ with domain $\mathbb{R}$ is many–one.

The function $f(x) = x^2$ with domain $x > 0$ is one–one.

The function $f(x) = \sin x$ with domain $0° \leqslant x \leqslant 360°$ is many–one.

The function $f(x) = \sin x$ with domain $-90° \leqslant x \leqslant 90°$ is one–one.

> These examples show the importance of including the domain of the function in its definition.

Composition of functions

Core 3

The diagram shows how two functions f and g may be combined.

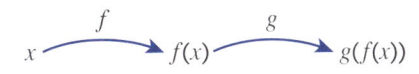

$$x \xrightarrow{f} f(x) \xrightarrow{g} g(f(x))$$

In this case, f is applied first to some value x giving $f(x)$. Then g is applied to the value $f(x)$ to give $g(f(x))$. This is usually written as $gf(x)$ and gf can be thought of as a new **composite function** defined from the functions f and g.

For example, if $f(x) = 3x$, $x \in \mathbb{R}$ and $g(x) = x + 2$, $x \in \mathbb{R}$ then
$gf(x) = g(3x) = 3x + 2$.

The order in which the functions are applied is important. The composite function fg is found by applying g first and then f.

In this case, $fg(x) = f(x + 2) = 3x + 6$.

Generally speaking, when two functions f and g are defined, the composite functions fg and gf will not be the same.

The inverse of a function

Core 3

The **inverse of a function** f is a function, usually written as f^{-1}, that *undoes* the effect of f. So the inverse of a function which adds 2 to every value, for example, will be a function that subtracts 2 from every value.

This can be written as $f(x) = x + 2$, $x \in \mathbb{R}$ and $f^{-1}(x) = x - 2$, $x \in \mathbb{R}$.

The domain of f^{-1} is given by the range of f.

Notice that $f^{-1}f(x) = f^{-1}(x + 2) = x$ and that $ff^{-1}(x) = f(x - 2) = x$.

A function can be either one–one or many–one, but only functions that are one–one can have an inverse. The reason is, that reversing a many–one function would give a mapping that is one–many, and this cannot be a function.

The diagram shows a many–one function. It does not have an inverse.

> Every value in the domain of a function can have only one image. This is an important property of functions.

> If you reverse this diagram then p would have more than one image.

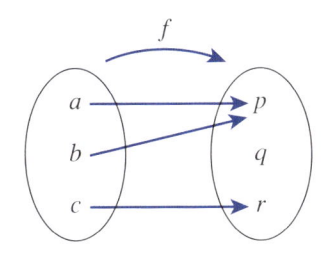

19

Core 3 and Core 4 (Pure Mathematics)

> Some important examples of restricting the domain of a function so that it will have an inverse are found in trigonometry. See page 28.

You can turn a many–one function into a one–one function by restricting its domain.

For example, the function $f(x) = x^2$, $x \in \mathbb{R}$ is many–one and so it does not have an inverse. However, the function $f(x) = x^2$, $x > 0$ is one–one and $f^{-1}(x) = \sqrt{x}$, $x > 0$.

Finding the inverse of a one–one function

Core 3

One way to find the inverse of a one–one function f is to write $y = f(x)$ and then rearrange this to make x the subject so that $x = f^{-1}(y)$. The inverse function is then usually defined in terms of x to give $f^{-1}(x)$. The domain of f^{-1} is the range of f.

> You can turn $f^{-1}(y)$ into $f^{-1}(x)$ by replacing every y with an x.

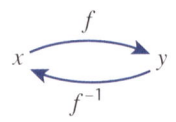

Example Find the inverse of the function $f(x) = \dfrac{x+2}{x-3}$, $x \neq 3$.

> Start by writing $y = f(x)$.

Define $y = \dfrac{x+2}{x-3}$

> Rearrange to find x.

then $y(x-3) = x+2$ — Multiply out the brackets.

$xy - 3y = x + 2$

> Collect the x terms on one side of the equation.

$xy - x = 3y + 2$

$x(y-1) = 3y + 2$

$x = \dfrac{3y+2}{y-1}$.

> $x = f^{-1}(y)$ so this defines the inverse function in terms of y.

> It is usual to express the inverse function in terms of x.

So $f^{-1}(x) = \dfrac{3x+2}{x-1}$, $x \neq 1$.

> The denominator cannot be allowed to be zero so 1 is not in the domain of f^{-1}.

Here are the graphs of $y = \dfrac{x+2}{x-3}$ and its inverse $y = \dfrac{3x+2}{x-1}$.

> The graph of the inverse function is the reflection of the original graph in the line $y = x$. This is true for any inverse function but the same scale must be used on both axes or the effect is distorted.

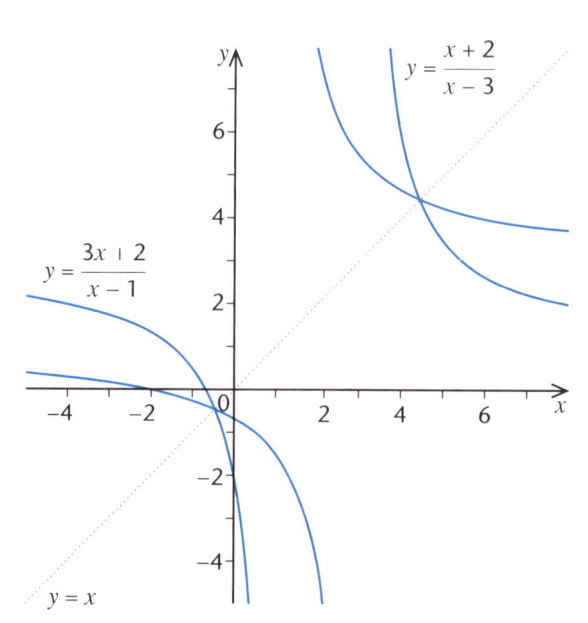

Core 3 and Core 4 (Pure Mathematics)

The modulus function

Core 3

The notation $|x|$ is used to stand for the modulus of x. This is defined as

$$|x| = \begin{cases} x & \text{when } x \geq 0 \quad \text{(when } x \text{ is positive, } |x| \text{ is just the same as } x\text{).} \\ -x & \text{when } x < 0 \quad \text{(when } x \text{ is negative, } |x| \text{ is the same as } -x\text{).} \end{cases}$$

The expression $|x-a|$ can be interpreted as the distance between the numbers x and a on the number line. In this way, the statement $|x-a| < b$ means that the distance between x and a is less than b.

It follows that $a - b < x < a + b$.

The graph of $y = |x|$

$|x|$ is never negative so the graph of $y = |x|$ doesn't go below the x-axis anywhere.

It follows that the graph of $y = |x|$ is the same as the graph of $y = x$ for positive values of x. But, when x is negative, the corresponding part of the graph of $y = x$ must be reflected in the x-axis to give the graph of $y = |x|$.

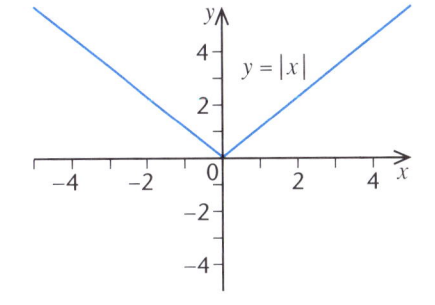

The graph of $y = |f(x)|$

The graph of $y = 1/x$ goes below the x-axis for negative values of x.

The graph of $y = |f(x)|$ is the same as the graph of $y = f(x)$ for positive values of $f(x)$. But, when $f(x)$ is negative, the corresponding part of the graph of $y = f(x)$ must be reflected in the x-axis to give the graph of $y = |f(x)|$.

The diagram shows the graph of $y = \left|\frac{1}{x}\right|$.

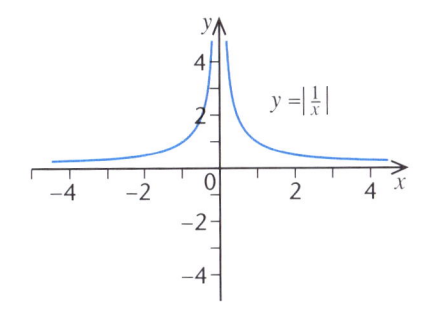

Transforming graphs

Core 3

The graph of some new function can often be obtained from the graph of a known function by applying a transformation. A summary of the standard transformations is given in the table.

You may need to apply a combination of transformations in some cases.

Known function	New function	Transformation
$y = f(x)$	$y = f(x) + a$	Translation through a units parallel to y-axis.
	$y = f(x - a)$	Translation through a units parallel to x-axis.
	$y = af(x)$	One-way stretch with scale factor a parallel to the y-axis.
	$y = f(ax)$	One-way stretch with scale factor $\frac{1}{a}$ parallel to the x-axis.

Core 3 and Core 4 (Pure Mathematics)

Examples

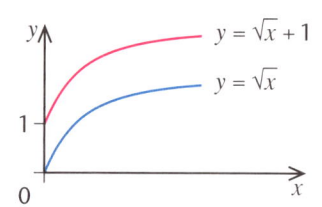

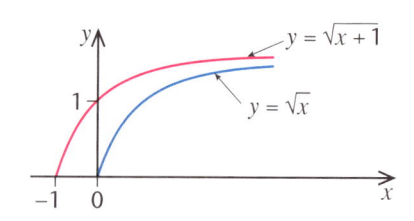

y = af(x) does not move points on the x-axis.

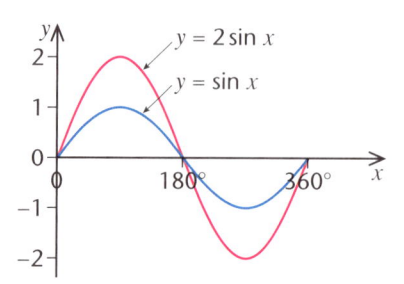

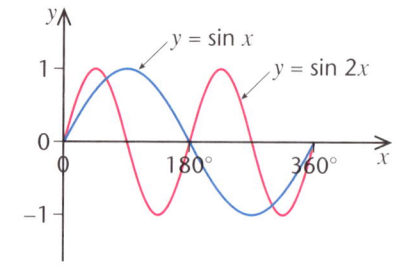

Sometimes you may have to perform more than one transformation.

Example

Describe how to transform the circle $x^2 + y^2 = 25$, with centre (0, 0) and radius 5, to the circle with centre (3, −8) and radius 5. Write down the new equation.

Key points from AS
- Circles *Revise AS page 25*

Translate 3 units parallel to the x-axis (to the right) and −8 units parallel to the y-axis (down).

$$x^2 + y^2 = 25 \rightarrow (x-3)^2 + (y+8)^2 = 25$$

The exponential function e^x

Core 3

Exponential functions are often used to represent, or model, patterns of:
- growth when $a > 1$
- decay when $a < 1$.

An **exponential function** is one where the variable is a power or exponent. For example, any function of the form $f(x) = a^x$, where a is a constant, is an exponential function.

The diagram shows some graphs of exponential functions for different values of a.

All of the graphs pass through the point (0, 1) and each one has a different gradient at this point. The value of the gradient depends on a.

Key points from AS
- Exponential functions *Revise AS page 36*

All scientific calculators will display values of e^x and 10^x. To display the value of e, for example, find e^1.

There is a particular value of a for which the curve has gradient 1 at (0, 1). The letter e is used to stand for this value and the function $f(x) = e^x$ is often called **the exponential function** because of its special importance to the subject.

e = 2.718281828... It is an irrational number and so it cannot be written exactly.

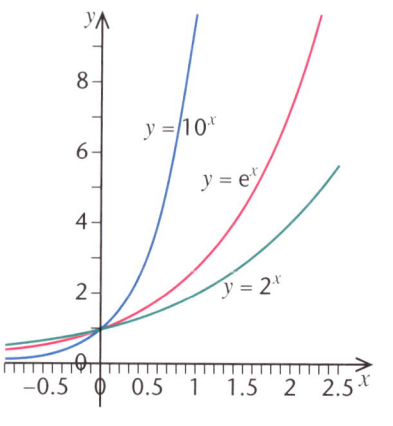

22

You may be asked to sketch other curves relating to e^x, for example

Notice $e^x > 0$ for all values of x.

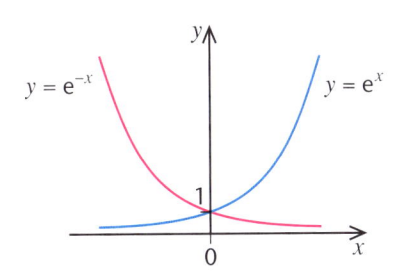

Reflection in y-axis.

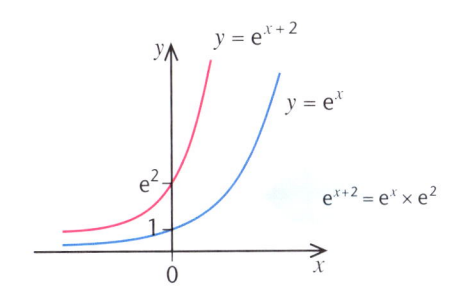

Translation 2 units left or stretch, factor e^2, parallel to y-axis.

$e^{x+2} = e^x \times e^2$

The natural logarithmic function, ln x

Core 3

Key points from AS

- **Logarithms**
  Revise AS page 36

Your calculator has a key for natural logs, probably labelled ln.

This is true for logs of any base.

If $y = e^x$ then $x = \log_e y$. This is usually written as $x = \ln y$.

Logarithms to the base e are called *natural logarithms*.

$e^0 = 1 \Rightarrow \ln 1 = 0$
$e^1 = e \Rightarrow \ln e = 1$

In fact, $\ln e^{f(x)} = f(x)$, for example $\ln e^{x^2} = x^2$, $\ln e^{\sin x} = \sin x$.

Since the exponential and natural logarithm functions are inverses of each other, their graphs are symmetrical about the line $y = x$.

The diagram shows that the domain of the natural logarithm function is given by $x > 0$.

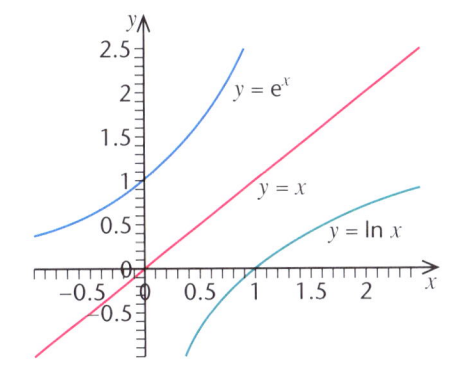

You may be asked to sketch other curves relating to $y = \ln x$, for example

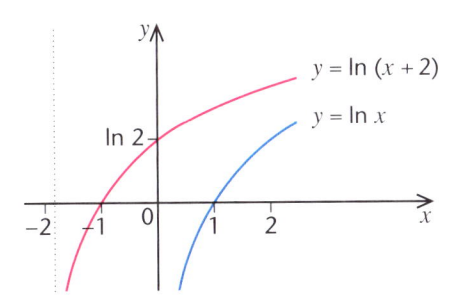

Translation 2 units to the left.

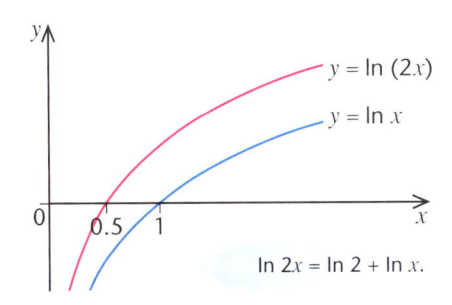

Stretch parallel to the x-axis factor $\frac{1}{2}$ or translation ln 2 units up.

$\ln 2x = \ln 2 + \ln x$.

23

Core 3 and Core 4 (Pure Mathematics)

Progress check

Core 3

1. $f(x) = x^2 + 5$ and $g(x) = 2x + 3$.
   - (a) Write $fg(x)$ in terms of x.
   - (b) Find $fg(10)$.
   - (c) Find the values of x for which $fg(x) = gf(x)$.

Core 3

2. Find the inverse of the function $f(x) = \dfrac{x+5}{x-2}$, $x \in \mathbb{R}$, $x \neq 2$ and state the domain of the inverse function.

Core 3

3. (a) Sketch $y = (x-2)(x+2)$.
   (b) Sketch $y = |x^2 - 4|$.

Core 3

4. Solve $|x - 2| < 5$.

Core 3

5. The diagram shows $y = f(x)$.
   On separate diagrams, sketch
   - (a) $y = f(x) + 1$
   - (b) $y = f(x + 1)$
   - (c) $y = f(2x)$
   - (d) $y = 2f(x)$
   - (e) $y = f(\tfrac{1}{2}x)$

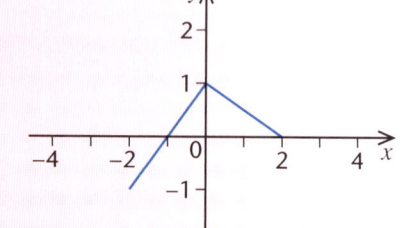

Core 3

6. On the same diagram, sketch $y = x^2$, $y = (x-2)^2$, $y = -x^2$.

Core 3

7. On the same diagram, sketch $y = e^x$, $y = e^{2x}$, $y = -e^x$.

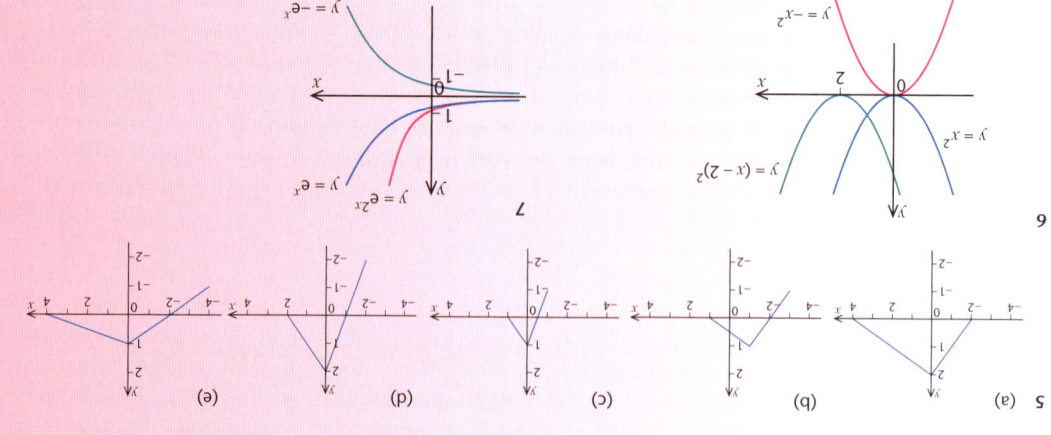

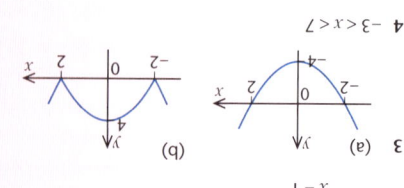

Answers

1. (a) $fg(x) = 4x^2 + 12x + 14$ (b) $fg(10) = 534$ (c) $x = -0.08452, -5.915$.

2. $f(x) = \dfrac{2x+5}{x-1}$, $x \in \mathbb{R}$, $x \neq 1$

3. (a), (b) [sketches]

4. $-3 < x < 7$

5. (a), (b), (c), (d), (e) [sketches]

6. [sketch: $y = x^2$, $y = (x-2)^2$, $y = -x^2$]

7. [sketch: $y = e^x$, $y = e^{2x}$, $y = -e^x$]

Core 3 and Core 4 (Pure Mathematics)

1.3 Coordinate geometry

After studying this section you should be able to:

- sketch curves in Cartesian and parametric form
- convert the equation of a curve between Cartesian and parametric form

LEARNING SUMMARY

Cartesian and parametric form

Core 4

Cartesian form

A curve in Cartesian form is defined in terms of x and y only, **for example**
$y = x^2 + 3x$ in which y is given **explicitly** in terms of x.
$x^2 + 2xy + y^2 = 9$ in which the equation is given **implicitly**.

Key points from AS

- Curve sketching
  Revise AS page 29

If all the powers of x are even, the graph is symmetrical in the y-axis.

Look for values that make the denominator zero.

To sketch a function given in Cartesian form:

1. Find where the curve **crosses the axes** by putting $x = 0$ and $y = 0$.
2. Check for **symmetry**.
3. Check for **stationary points**.
4. Check what happens for **large values of x and y**.
5. Check for **discontinuities**.

Find when $\dfrac{dy}{dx} = 0$ and then investigate.

Parametric form

The parameter θ is often used when trigonometric functions are involved.

A curve is defined in **parametric form** by expressing x and y in terms of a third variable, for example $x = f(t)$, $y = g(t)$. In this case, t is the parameter.

Example

The curve is plotted by finding the values of x and y for various values of the parameter.

Plot the curve $x = 2t^2$, $y = 4t$ for $-3 \leq t \leq 3$ and find the Cartesian equation of the curve.

The curve is a parabola.

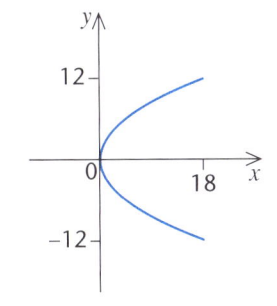

t	−3	−2	−1	0	1	2	3
x	18	8	2	0	2	8	18
y	−12	−8	−4	0	4	8	12

Eliminate t to obtain the Cartesian equation.

$y = 4t \implies t = \dfrac{y}{4}$.

Substituting into $x = 2t^2$ gives $x = 2\left(\dfrac{y}{4}\right)^2 = \dfrac{y^2}{8}$ so $y^2 = 8x$.

Key points from AS

- Trigonometric identities
  Revise AS page 47

When the parameter is θ, it may be necessary to use a trigonometric identity to find the Cartesian equation.

25

Key points from AS

- **Circle**
  Revise AS page 25

For example, when $x = a\cos\theta$ and $y = a\sin\theta$, use $\cos^2\theta + \sin^2\theta = 1$. This gives $\frac{x^2}{a^2} + \frac{y^2}{a^2} = 1$ or $x^2 + y^2 = a^2$.

This is a circle, centre (0, 0), radius a.

Also, when $x = a\cos\theta$, $y = b\sin\theta$, $\frac{x^2}{a^2} + \frac{y^2}{b^2} = 1$.

This curve is an ellipse.

Progress check

Core 4

1 Sketch the curve $y = x^3 + 3x^2$.

Core 4

2 A curve is defined parametrically by $x = 2t$, $y = \frac{2}{t}$.

(a) Give the Cartesian equation of the curve.

(b) Sketch the curve.

Core 4

3 Describe the curve defined parametrically by $x = 5\cos\theta$, $y = 5\sin\theta$ and give its Cartesian equation.

1

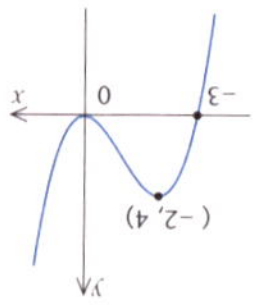

2 (a) $xy = 4$ (hyperbola)

(b)

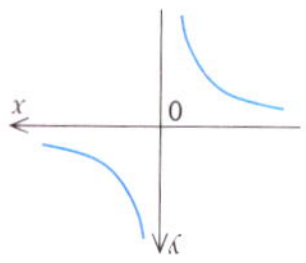

3 Circle, centre (0, 0), radius 5; $x^2 + y^2 = 25$.

Core 3 and Core 4 (Pure Mathematics)

1.4 Trigonometry

After studying this section you should be able to:

- understand the functions secant, cosecant and cotangent and be able to draw their graphs
- understand and use the Pythagorean identities
- understand and use the compound angle formulae for sine, cosine and tangent
- use identities to solve equations

Secant, cosecant and cotangent

Core 3

These trigonometric functions, commonly known as **sec**, **cosec** and **cot**, are defined from the more familiar sin, cos and tan functions as follows:

$$\sec x = \frac{1}{\cos x} \qquad \cosec x = \frac{1}{\sin x} \qquad \cot x = \frac{1}{\tan x}$$

Key points from AS
- **Trigonometry** pages 44–48

Each function is periodic and has either line or rotational symmetry. The domain of each one must be restricted to avoid division by zero.
You need to be able to work in degrees or radians.
Remember $180° = \pi$ radians.

$y = \sec x$

- The period of sec is 360° (2π radians) to match the period of cos.
- Notice that $\sec x$ is undefined whenever $\cos x = 0$.
- The graph is symmetrical about every vertical line passing through a vertex.
- It has rotational symmetry of order 2 about the points in the x-axis corresponding to $90° \pm 180°n$ ($\frac{\pi}{2} \pm \pi n$).

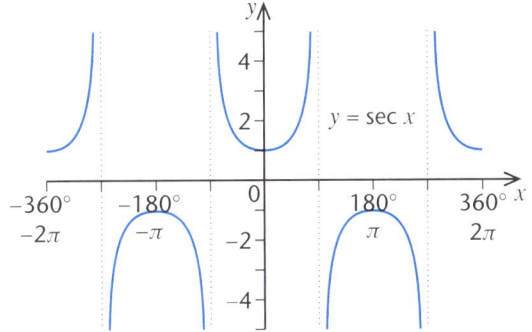

$y = \cosec x$

- The period of cosec is 360° (2π radians) to match the period of sin.
- Notice that $\cosec x$ is undefined whenever $\sin x = 0$.
- The graph is symmetrical about every vertical line passing through a vertex.
- It has rotational symmetry of order 2 about the points on the x-axis corresponding to $0°, \pm 180°, \pm 360°, \ldots$, $(0, \pm \pi, \pm 2\pi, \ldots)$.

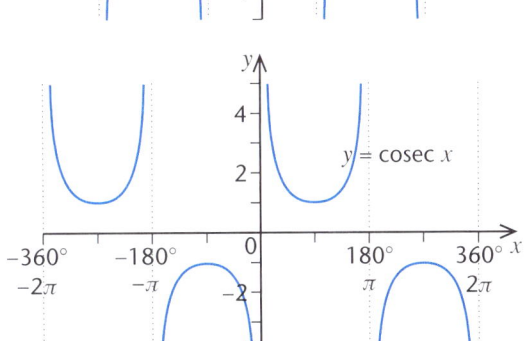

$y = \cot x$

$\frac{1}{\tan x} = \frac{\cos x}{\sin x}$.

- The period of cot is 180° to match the period of tan.
- Notice that $\cot x$ is undefined whenever $\sin x = 0$.
- The graph has rotational symmetry of order 2 about the points on the x-axis corresponding to $0°, \pm 90°, \pm 180°, \ldots$, $(0, \pm \frac{\pi}{2}, \pm \pi, \ldots)$.

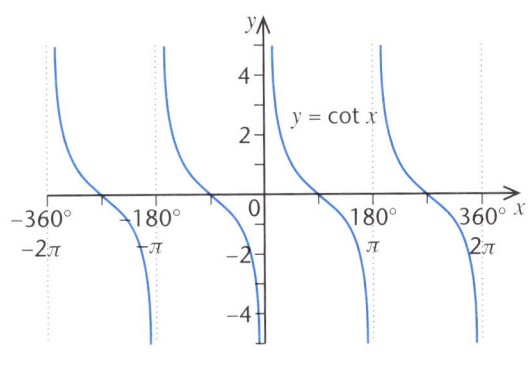

Core 3 and Core 4 (Pure Mathematics)

Inverse trigonometric functions

Core 3

The sine, cosine and tangent functions are all many–one and so do not have inverses on their full domains. However, it is possible to restrict their domains so that each one has an inverse. The graphs of these inverse functions are given below.

See page 19 to find out about inverse functions.

You need to know these graphs and to understand the need for restricting the domains.

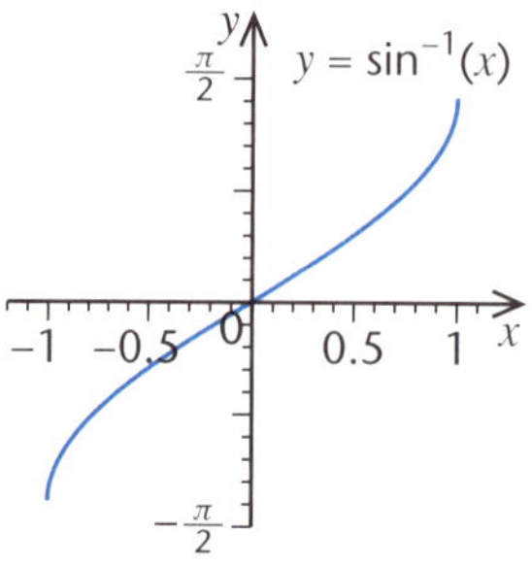

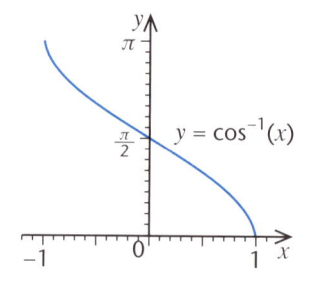

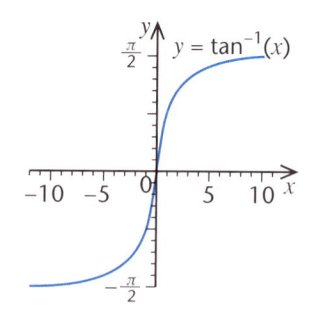

$f(x) = \sin x, -\frac{\pi}{2} \leq x \leq \frac{\pi}{2}$
$(-90° \leq x \leq 90°)$
$\Rightarrow f^{-1}(x) = \sin^{-1} x, -1 \leq x \leq 1.$

$f(x) = \cos x, 0 \leq x \leq \pi$
$(0° \leq x \leq 180°)$
$\Rightarrow f^{-1}(x) = \cos^{-1} x, -1 \leq x \leq 1.$

$f(x) = \tan x, -\frac{\pi}{2} < x < \frac{\pi}{2}$
$(-90° < x < 90°)$
$\Rightarrow f^{-1}(x) = \tan^{-1} x, x \in \mathbb{R}.$

This is the notation used for inverse trig' functions on your calculator. But some texts use:
arcsin for $\sin^{-1}$
arccos for $\cos^{-1}$
arctan for $\tan^{-1}$.

$\sin^{-1} x$ means:
'the angle whose sine is x'
e.g. $\sin^{-1} 0.5 = 30°$.

In radians $\sin^{-1} 0.5 = \frac{\pi}{6}$.

$\cos^{-1} x$ means:
'the angle whose cosine is x'
e.g. $\cos^{-1} 0.5 = 60°$.

In radians $\cos^{-1} 0.5 = \frac{\pi}{3}$.

$\tan^{-1} x$ means:
'the angle whose tangent is x'
e.g. $\tan^{-1} 1 = 45°$.

In radians $\tan^{-1} 1 = \frac{\pi}{4}$.

When you use the functions $\sin^{-1}$, $\cos^{-1}$ and $\tan^{-1}$ on your calculator, the value given is called the **principal value** (PV).

Trigonometric identities

Core 3

You should *learn* the identities listed below. It is so much easier to simplify expressions, establish new identities and solve equations if you have a firm grasp of these basic results.

The second result can be obtained from the first by dividing throughout by $\cos^2 \theta$. The third result can be obtained in a similar way, by dividing throughout by $\sin^2 \theta$.

- The **Pythagorean identities** are:

    $\cos^2 \theta + \sin^2 \theta \equiv 1$

    $1 + \tan^2 \theta \equiv \sec^2 \theta$

    $\cot^2 \theta + 1 \equiv \csc^2 \theta$

- The **compound angle identities** for $(A + B)$ are:

    $\sin(A + B) \equiv \sin A \cos B + \cos A \sin B$

    $\cos(A + B) \equiv \cos A \cos B - \sin A \sin B$

    $\tan(A + B) \equiv \dfrac{\tan A + \tan B}{1 - \tan A \tan B}.$

    The corresponding results for $(A - B)$ are:

These results can be worked out from the ones for $(A + B)$. Just replace every '+' with a '–' and vice versa. This makes it easier to learn the results.

$\sin(A - B) \equiv \sin A \cos B - \cos A \sin B$

$\cos(A - B) \equiv \cos A \cos B + \sin A \sin B$

$\tan(A - B) \equiv \dfrac{\tan A - \tan B}{1 + \tan A \tan B}.$

Core 3 and Core 4 (Pure Mathematics)

> You can obtain these results from the compound angle formulae by setting $A = B$.

- The **double angle identities** are:
$$\sin 2A \equiv 2 \sin A \cos A$$
$$\cos 2A \equiv \cos^2 A - \sin^2 A$$
$$\equiv 2\cos^2 A - 1$$
$$\equiv 1 - 2\sin^2 A$$

> These variations are very useful – it is worth knowing them.

$$\tan 2A \equiv \frac{2 \tan A}{1 - \tan^2 A}.$$

- The **half angle identities** are:

> These results will be useful for integration later in the course.

$$\sin^2 \tfrac{1}{2}A \equiv \tfrac{1}{2}(1 - \cos A)$$
$$\cos^2 \tfrac{1}{2}A \equiv \tfrac{1}{2}(1 + \cos A).$$

Proving identities

Core 3

The basic technique for proving a given identity is:

- Start with the expression on one side of the identity.
- Use known identities to replace some part of it with an equivalent form.
- Simplify the result and compare with the expression on the other side.

> This is a skill. The more you do, the better you will become.

With practice you get to know which parts to replace so that the required result is established.

Example

Prove that $\dfrac{\cos 2A}{\cos A - \sin A} \equiv \cos A + \sin A.$

Starting with the LHS:

$$\frac{\cos 2A}{\cos A - \sin A} \equiv \frac{\cos^2 A - \sin^2 A}{\cos A - \sin A}$$ Using one of the double angle results.

$$\equiv \frac{(\cos A + \sin A)(\cos A - \sin A)}{\cos A - \sin A}$$ Using the difference of two squares.

$$\equiv \cos A + \sin A \equiv \text{RHS}.$$ Cancelling the common factor.

It is often a good idea to replace sec, cosec and cot with sin, cos and tan at the start.

Example

Prove that $\sec^2 A \equiv \dfrac{\operatorname{cosec} A}{\operatorname{cosec} A - \sin A}.$

> It's often easier to start with the more complicated looking side and try to simplify it.

$$\text{RHS} \equiv \frac{\frac{1}{\sin A}}{\frac{1}{\sin A} - \sin A} \equiv \frac{1}{1 - \sin^2 A}$$ Replacing $\operatorname{cosec} A$ with $\dfrac{1}{\sin A}$ and multiplying the numerator and denominator by $\sin A$.

$$\equiv \frac{1}{\cos^2 A} \equiv \sec^2 A \equiv \text{LHS}.$$ Using one of the Pythagorean identities and the definition of $\sec A$.

Core 3 and Core 4 (Pure Mathematics)

Trigonometric equations

Core 3

Identities may be used to re-write an equation in a form that is easier to solve.

Example Solve the equation $\cos 2x + \sin x + 1 = 0$ for $-\pi \leqslant x \leqslant \pi$.

Replacing $\cos 2x$ with $1 - 2\sin^2 x$ gives $1 - 2\sin^2 x + \sin x + 1 = 0$

$$\Rightarrow 2\sin^2 x - \sin x - 2 = 0.$$

You need to recognise that this is a quadratic in $\sin x$.

Using the quadratic formula gives $\sin x = \dfrac{1 \pm \sqrt{1+16}}{4} = \dfrac{1 \pm \sqrt{17}}{4}$

$$\Rightarrow \sin x = 1.28 \ldots \text{ (no solutions) or } \sin x = -0.78077 \ldots.$$

Using $\sin^{-1}$ gives $x = -0.8959$ radians to 4 d.p.

Set calculator to radian mode.

$-1 \leqslant \sin x \leqslant 1$ for all values of x.

Another solution in the interval $-\pi \leqslant x \leqslant \pi$ is $-\pi + 0.8959$ giving $x = -2.2457$ radians to 4 d.p.

The diagram shows there are no other solutions in this interval.

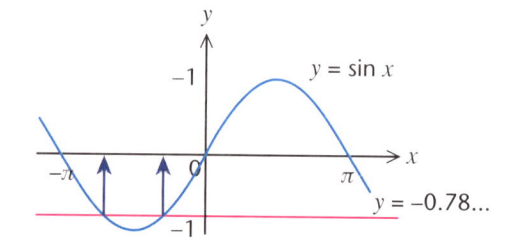

> **KEY POINT**
> An important result is that the expression $a \cos\theta + b \sin\theta$ can be written in any of the equivalent forms $r\cos(\theta \pm \alpha)$ or $r\sin(\theta \pm \alpha)$ for a suitable choice of r and α. The convention is to take $r > 0$.

One application of the result is to solve equations of the form $a\cos\theta + b\sin\theta = c$.

Example Express $3\cos\theta - 4\sin\theta$ in the form $r\cos(\theta + \alpha)$, $r > 0$. Use the result to solve the equation $3\cos\theta - 4\sin\theta = 2$ for $-\pi \leqslant \theta \leqslant \pi$.

You need to expand $r\cos(\theta + \alpha)$ and match corresponding parts of the identity.

$r\cos(\theta + \alpha) \equiv 3\cos\theta - 4\sin\theta$

$\Rightarrow r\cos\theta\cos\alpha - r\sin\theta\sin\alpha \equiv 3\cos\theta - 4\sin\theta$

$\Rightarrow r\cos\alpha = 3$ and $r\sin\alpha = 4$

$\dfrac{r\sin\alpha}{r\cos\alpha} = \dfrac{4}{3}$

$\Rightarrow \tan\alpha = \dfrac{4}{3}$

$\Rightarrow r = \sqrt{3^2 + 4^2} = 5$ and $\alpha = \tan^{-1}\left(\dfrac{4}{3}\right) = 0.9273^c$ to 4 d.p.

$\Rightarrow 3\cos\theta - 4\sin\theta \equiv 5\cos(\theta + 0.9273^c)$.

Using this result in the equation $3\cos\theta - 4\sin\theta = 2$ gives

$$5\cos(\theta + 0.9273^c) = 2$$
$$\Rightarrow \cos(\theta + 0.9273^c) = 0.4$$

This gives $\theta + 0.9273^c = \pm 1.1593^c \Rightarrow \theta = 0.232^c$ or $\theta = -2.087^c$ to 3 d.p.

Progress check

Core 3
1. Prove the identity $\sin 2\theta \equiv \dfrac{2\tan\theta}{1 + \tan^2\theta}$.

Core 3
2. Solve the equation $\cos 2x - \sin x = 0$ for $0° \leqslant x \leqslant 360°$.

Core 3
3. (a) Write $5\sin x + 12\cos x$ in the form $r\sin(x + \alpha)$ where $0 < \alpha < \dfrac{\pi}{2}$.

   (b) Solve the equation $5\sin x + 12\cos x = 7$ for $-\pi \leqslant x \leqslant \pi$.

3 (a) $13\sin(x + 1.176^c)$ (b) $x = -0.607^c$ or $x = 1.397^c$ to 3 d.p.
2. $x = 30°, 150°$ or $270°$
1. RHS $\equiv \dfrac{2\tan\theta}{\sec^2\theta} \equiv 2\tan\theta \times \cos^2\theta \equiv 2\sin\theta\cos\theta \equiv \sin 2\theta \equiv$ LHS.

Core 3 and Core 4 (Pure Mathematics)

1.5 Differentiation

LEARNING SUMMARY

After studying this section you should be able to:

- use the chain rule to differentiate composite functions
- differentiate functions based on e^x and $\ln x$
- differentiate trigonometric functions
- differentiate using the product and quotient rules
- understand and use the relation $\dfrac{dy}{dx} = 1 \div \dfrac{dx}{dy}$
- find derivatives of functions defined parametrically and implicitly

The chain rule

Core 3

> **KEY POINT**
>
> The **chain rule** can be written as $\dfrac{dy}{dx} = \dfrac{dy}{du} \times \dfrac{du}{dx}$, where u is a function of x.

It shows how to differentiate a composite function by separating the process into simpler stages and combining the results.

Example Differentiate (a) $y = (3x + 2)^{10}$

(b) $y = \sqrt{x^2 - 9}$.

Key points from AS
- Differentiation
  Revise AS pages 28–31

(a) Taking $u = 3x + 2$ gives $y = u^{10}$ and $\dfrac{dy}{du} = 10u^9$, so $\dfrac{dy}{du} = 10(3x + 2)^9$.

It also gives $\dfrac{du}{dx} = 3$.

So, from the chain rule: $\dfrac{dy}{dx} = 10(3x + 2)^9 \times 3 = 30(3x + 2)^9$.

(b) Taking $u = x^2 - 9$ gives $y = \sqrt{u} = u^{\frac{1}{2}}$, and $\dfrac{dy}{du} = \tfrac{1}{2} u^{-\frac{1}{2}}$, so $\dfrac{dy}{du} = \tfrac{1}{2}(x^2 - 9)^{-\frac{1}{2}}$.

It also gives $\dfrac{du}{dx} = 2x$.

So, from the chain rule: $\dfrac{dy}{dx} = \tfrac{1}{2}(x^2 - 9)^{-\frac{1}{2}} \times 2x = \dfrac{x}{\sqrt{x^2 - 9}}$.

The chain rule can also be used to establish results for **connected rates of change**.

For example, the rate of change of the volume of a sphere can be written as $\dfrac{dv}{dt}$.

The corresponding rate of change of the radius of the sphere can be written as $\dfrac{dr}{dt}$. Using the chain rule, the connection between these rates of change is given by:

$$\dfrac{dv}{dt} = \dfrac{dv}{dr} \times \dfrac{dr}{dt}.$$

Also, since $v = \tfrac{4}{3} \pi r^3$, it follows that $\dfrac{dv}{dr} = 4\pi r^2$, so, the connection can now be written as $\dfrac{dv}{dt} = 4\pi r^2 \dfrac{dr}{dt}$.

Core 3 and Core 4 (Pure Mathematics)

The exponential function e^x

Core 3

The exponential function was defined on page 22. It has the special property that it is unchanged by differentiation. So, if $y = e^x$ then $\frac{dy}{dx} = e^x$.

Constant multiples of the exponential function behave in the same way, so if $y = ae^x$ then $\frac{dy}{dx} = ae^x$ where a is any constant.

Examples
Differentiate with respect to x:

(a) $y = 2e^x$ (b) $y = x^3 + e^x$ (c) $f(x) = \frac{1}{x} - 3e^x$

(a) $\frac{dy}{dx} = 2e^x$ (b) $\frac{dy}{dx} = 3x^2 + e^x$ (c) $f'(x) = -\frac{1}{x^2} - 3e^x$

You may need to use the chain rule.

Examples
Differentiate with respect to x:

(a) $y = e^{3x+2}$ (b) $y = e^{4x^2}$

(a) Let $u = 3x + 2$ so that $y = e^u$

$\frac{du}{dx} = 3$ and $\frac{dy}{du} = e^u$

Using the chain rule

$\frac{dy}{dx} = \frac{dy}{du} \times \frac{du}{dx} = e^u \times 3 = 3e^{3x+2}$

(b) Let $u = 4x^2$, so that $y = e^u$

$\frac{du}{dx} = 8x$ and $\frac{dy}{du} = e^u$

$\frac{dy}{dx} = \frac{dy}{du} \times \frac{du}{dx} = e^u \times 8x = 8xe^{4x^2}$

In general, it can be shown using the chain rule

> **KEY POINT**
> If $y = e^{f(x)}$ then $\frac{dy}{dx} = f'(x)e^{f(x)}$.

$f'(x)$ is the derivative, with respect to x, of $f(x)$.

In particular, it is useful to remember that if $y = e^{ax+b}$, then $\frac{dy}{dx} = ae^{ax+b}$

Logarithmic functions $\ln x$ and $\ln f(x)$

Core 3

> **KEY POINT**
> If $y = \ln x$ then $\frac{dy}{dx} = \frac{1}{x}$.

Core 3 and Core 4 (Pure Mathematics)

Key points from AS

- Logarithms
  Revise AS page 36

$\log a + \log b = \log(ab)$.
$\log x^k = k \log x$.

This result is easily extended to functions of the form $y = \ln(kx)$ and $y = \ln(x^k)$ using the laws of logarithms.

If $y = \ln(kx)$, then $y = \ln k + \ln x$

giving $\dfrac{dy}{dx} = \dfrac{1}{x} + 0 = \dfrac{1}{x}$.

> Since $\ln k$ is a constant, its derivative is zero.

For example, if $y = \ln(5x)$, then $\dfrac{dy}{dx} = \dfrac{1}{x}$.

If $y = \ln(x^k)$, then $y = k \ln x$ giving $\dfrac{dy}{dx} = k \times \dfrac{1}{x} = \dfrac{k}{x}$.

For example, if $y = \ln\left(\dfrac{1}{x^2}\right) = \ln(x^{-2}) = -2 \ln x$, then $\dfrac{dy}{dx} = -2 \times \dfrac{1}{x} = -\dfrac{2}{x}$.

Always check whether a logarithmic function can be simplified before attempting to differentiate.

You may need to use the chain rule to differentiate $y = \ln f(x)$.

Example

Differentiate $y = \ln(x^2 + 3x)$.

> This function cannot be simplified.

Let $u = x^2 + 3x$ so that $y = \ln u$

$\dfrac{du}{dx} = 2x + 3$ and $\dfrac{dy}{du} = \dfrac{1}{u} = \dfrac{1}{x^2 + 3x}$

Using the chain rule
$\dfrac{dy}{dx} = \dfrac{dy}{du} \times \dfrac{du}{dx} = \dfrac{1}{x^2 + 3x} \times (2x + 3)$

$= \dfrac{2x + 3}{x^2 + 3x}$

> Notice that $2x + 3$ is the derivative of $x^2 + 3x$.

In general, it can be shown using the chain rule

> This can be used directly without showing the chain rule working.

KEY POINT

$$\dfrac{d}{dx} \ln(f(x)) = \dfrac{f'(x)}{f(x)}.$$

Progress check

Find $\dfrac{dy}{dx}$ in each of the following questions.

Core 3
1 $y = (4 - 3x)^6$

Core 3
2 $y = \sqrt{2x + 1}$

Core 3
3 $y = e^{3x+1} + 3x$

Core 3
4 $y = \ln(7x) + \ln(x^4)$

Core 3
5 $y = \ln(x^2 - 3)$

5. $\dfrac{2x}{x^2 - 3}$
4. $\dfrac{5}{x}$
3. $3e^{3x+1} + 3$
2. $(2x + 1)^{-\frac{1}{2}}$
1. $-18(4 - 3x)^5$

33

Core 3 and Core 4 (Pure Mathematics)

Trigonometric functions

Core 4

You should learn the derivatives of the three main **trigonometric functions**:

$$\frac{d}{dx}(\sin x) = \cos x \qquad \frac{d}{dx}\sin(ax+b) = a\cos(ax+b)$$

$$\frac{d}{dx}(\cos x) = -\sin x \qquad \frac{d}{dx}\cos(ax+b) = -a\sin(ax+b)$$

$$\frac{d}{dx}(\tan x) = \sec^2 x \qquad \frac{d}{dx}\tan(ax+b) = a\sec^2(ax+b)$$

Example

When differentiating, the angle must be in radians.

Find the exact value of the gradient of $y = \cos 2x$ when $x = \dfrac{\pi}{6}$.

$$y = \cos 2x \Rightarrow \frac{dy}{dx} = -2\sin 2x$$

When $x = \dfrac{\pi}{6}$, $\dfrac{dy}{dx} = -2\sin\dfrac{\pi}{3} = -2 \times \dfrac{\sqrt{3}}{2} = -\sqrt{3}$

The **chain rule** can often be used to differentiate expressions involving trigonometric functions.

Example

Differentiate: (a) $y = 5\sin^4 x$ (b) $y = e^{\tan 4x}$ (c) $y = \ln(\sin x)$

$\sin^4 x = (\sin x)^4$.

(a) Let $u = \sin x$ so that $y = 5u^4$

Then $\dfrac{du}{dx} = \cos x$ and $\dfrac{dy}{du} = 20u^3 = 20(\sin x)^3 = 20\sin^3 x$

With practice you should be able to do the working mentally.

By the chain rule:

$$\frac{dy}{dx} = \frac{dy}{du} \times \frac{du}{dx} = 20\sin^3 x \times (\cos x)$$

$$= 20\cos x \sin^3 x.$$

Although not essential, it is usual to write an expression with the least complicated factors first.

(b) Let $u = \tan 4x$ so that $y = e^u$

$$\frac{du}{dx} = 4\sec^2 4x \text{ and } \frac{dy}{du} = e^u = e^{\tan 4x}.$$

$$\therefore \frac{dy}{dx} = e^{\tan 4x} \times 4\sec^2 4x$$

$$= 4\sec^2 4x\, e^{\tan 4x}.$$

In general, by the chain rule:

$$\frac{d}{dx}(e^{f(x)}) = f'(x)e^{f(x)}.$$

(c) Use the result $y = \ln f(x) \Rightarrow \dfrac{dy}{dx} = \dfrac{f'(x)}{f(x)}$

$$y = \ln(\sin x)$$

$$\frac{dy}{dx} = \frac{\cos x}{\sin x}$$

$$= \cot x.$$

$f(x) = \sin x$, $f'(x) = \cos x$.

Core 3 and Core 4 (Pure Mathematics)

The product rule

Core 3

Expressions written as the product of two functions, $f(x) \times g(x)$, can be differentiated using the **product rule**.

> **KEY POINT**
> If $y = uv$, where $u = f(x)$ and $v = g(x)$, then $\dfrac{dy}{dx} = u\dfrac{dv}{dx} + v\dfrac{du}{dx}$.

Example
Differentiate with respect to x: (a) $y = x^2 \sin 4x$ (b) $y = x^3 e^{-x}$

Example (a) Core 4

(a) Let $y = uv$, where $u = x^2$ and $v = \sin 4x$ Side working

$$\dfrac{dy}{dx} = u\dfrac{dv}{dx} + v\dfrac{du}{dx}$$

$u = x^2 \Rightarrow \dfrac{du}{dx} = 2x$

With practice, the side working can be done mentally.

$$= x^2 \times 4\cos 4x + \sin 4x \times 2x$$

$v = \sin 4x \Rightarrow \dfrac{dv}{dx} = 4\cos 4x$

$$= 2x(2x \cos 4x + \sin 4x)$$

Tidy the answer by taking out factors if possible.

(b) Let $y = uv$, where $u = x^3$ and $v = e^{-x}$ Side working

$$\dfrac{dy}{dx} = x^3 \times (-e^{-x}) + e^{-x} \times 3x^2$$

$u = x^3 \Rightarrow \dfrac{du}{dx} = 3x^2$

$$= x^2 e^{-x}(-x + 3)$$

$v = e^{-x} \Rightarrow \dfrac{dv}{dx} = -e^{-x}$

The quotient rule

Core 3

Expressions written in the form $\dfrac{f(x)}{g(x)}$ can be differentiated using the **quotient rule**.

> **KEY POINT**
> If $y = \dfrac{u}{v}$ where $u = f(x)$ and $v = g(x)$, then $\dfrac{dy}{dx} = \dfrac{v\dfrac{du}{dx} - u\dfrac{dv}{dx}}{v^2}$.

Before using the quotient rule, check whether the expression can be simplified first, such as in
$y = \dfrac{4x^2 + 3x + 2}{x}$.

Example
Find $\dfrac{dy}{dx}$ when $y = \dfrac{e^{2x}}{x^3}$.

Side working:

$$\dfrac{dy}{dx} = \dfrac{v\dfrac{du}{dx} - u\dfrac{dv}{dx}}{v^2}$$

$u = e^{2x} \Rightarrow \dfrac{du}{dx} = 2e^{2x}$

$$= \dfrac{x^3(2e^{2x}) - e^{2x}(3x^2)}{(x^3)^2}$$

$v = x^3 \Rightarrow \dfrac{dv}{dx} = 3x^2$

Take out factors.

$$= \dfrac{x^2 e^{2x}(2x - 3)}{x^6}$$

Simplify.

$$= \dfrac{e^{2x}(2x - 3)}{x^4}$$

Core 3 and Core 4 (Pure Mathematics)

Example

Find the coordinates of the stationary point on the curve $y = \dfrac{\ln x}{x}$.

Side working:

$$\dfrac{dy}{dx} = \dfrac{x \times \dfrac{1}{x} - \ln x \times 1}{x^2}$$

$$u = \ln x \Rightarrow \dfrac{du}{dx} = \dfrac{1}{x}$$

$$= \dfrac{1 - \ln x}{x^2}$$

$$v = x \Rightarrow \dfrac{dv}{dx} = 1$$

$\dfrac{dy}{dx} = 0$ when $\ln x = 1$, i.e. when $x = e$

When $x = e$, $y = e^{-1}$, so the stationary point is at (e, e^{-1}).

Key points from AS
- Stationary points
  Revise AS page 29

$\ln e = 1$.

Alternative formats of product and quotient rules when $u = f(x)$ and $v = g(x)$:

Product rule: If $y = f(x) \times g(x)$, then $\dfrac{dy}{dx} = f'(x)g(x) + f(x)g'(x)$

Quotient rule: If $y = \dfrac{f(x)}{g(x)}$, then $\dfrac{dy}{dx} = \dfrac{f'(x)g(x) - f(x)g'(x)}{[g(x)]^2}$

These formats are sometimes quoted. Learn and use the ones you find easiest to remember.

Using the result $\dfrac{dy}{dx} = 1 \div \dfrac{dx}{dy}$

Core 3

> $\dfrac{dy}{dx} = 1 \div \dfrac{dx}{dy} = \dfrac{1}{\dfrac{dx}{dy}}$.
>
> **KEY POINT**

For example, if $x = 4y + 2$, then $\dfrac{dx}{dy} = 4$ and $\dfrac{dy}{dx} = \dfrac{1}{\dfrac{dx}{dy}} = \dfrac{1}{4}$.

Parametric functions

Core 4

> If x and y are each expressed in terms of a **parameter** t,
> then $\dfrac{dy}{dx} = \dfrac{dy}{dt} \times \dfrac{dt}{dx}$. Remember that $\dfrac{dt}{dx} = \dfrac{1}{dx/dt}$.
>
> **KEY POINT**

Example

A curve is defined parametrically by $x = 3t^2$, $y = 4t$.

Find $\dfrac{dy}{dx}$ when $y = 8$

$x = 3t^2 \Rightarrow \dfrac{dx}{dt} = 6t \Rightarrow \dfrac{dt}{dx} = \dfrac{1}{6t}$

$y = 4t \Rightarrow \dfrac{dy}{dt} = 4$

Core 3 and Core 4 (Pure Mathematics)

$$\therefore \frac{dy}{dx} = \frac{dy}{dt} \times \frac{dt}{dx} = 4 \times \frac{1}{6t} = \frac{2}{3t}$$

> Find the value of t when $y = 8$.

When $y = 8$, $4t = 8$, i.e. $t = 2$; when $t = 2$, $\frac{dy}{dx} = \frac{2}{3t} = \frac{2}{6} = \frac{1}{3}$.

Implicit functions

Core 4

> To find $\frac{dy}{dx}$ when an equation in x and y is given **implicitly**, differentiate each term with respect to x, remembering that $\frac{d}{dx}(f(y)) = \frac{d}{dy}f(y) \times \frac{dy}{dx}$.

Example

The point $P(2, 6)$ lies on the circle $x^2 + y^2 + 2x - 4y - 20 = 0$.

(a) Show that $\frac{dy}{dx} = \frac{x+1}{2-y}$

(b) Calculate the gradient of the tangent at P.

(a) $2x + 2y\frac{dy}{dx} + 2 - 4\frac{dy}{dx} + 0 = 0$

> Differentiate term by term with respect to x
> $\frac{d}{dx}(y) = \frac{dy}{dx}$.

$$x + y\frac{dy}{dx} + 1 - 2\frac{dy}{dx} = 0$$

> Get the terms in $\frac{dy}{dx}$ on one side of the equation and all the other terms on the other side.

> Divide throughout by 2 to simplify the equation.

$$x + 1 = 2\frac{dy}{dx} - y\frac{dy}{dx}$$

$$\Rightarrow x + 1 = \frac{dy}{dx}(2 - y)$$

> Take out $\frac{dy}{dx}$ as a factor.

$$\therefore \frac{dy}{dx} = \frac{x+1}{2-y}$$

(b) At P, $x = 2$ and $y = 6$,

so $\frac{dy}{dx} = \frac{3}{-4} = -\frac{3}{4}$.

> The value of $\frac{dy}{dx}$ when $x = 2$ gives the gradient of the tangent at (2, 6)

The gradient of the tangent at P is $-\frac{3}{4}$.

Core 3 and Core 4 (Pure Mathematics)

Progress check

Core 4

1. Differentiate with respect to x:
   (a) $y = 2 \sin 6x$ (b) $y = 4 \cos^3 x$ (c) $y = 2 \tan 3x$
   (d) $y = e^{\sin x}$ (e) $y = \sin(x^3)$

Core 4

2. If $y = \sin^2 x$, show that $\dfrac{dy}{dx} = \sin 2x$.   *Use a trigonometric identity.*

Core 3 (Part (c) Core 4)

3. Differentiate with respect to x:
   (a) $\ln(x^2 + 3)$ (b) $\ln(2x - 4)^5$ (c) $\ln(\cos 2x)$ (d) $\ln(6x - 1)$

4. Find $\dfrac{dy}{dx}$, simplifying your answers.

Core 4

   (a) $y = x^2 \cos 3x$
   (b) $y = e^{2x} \sin x$

Core 3

   (c) $y = x \ln x$
   (d) $y = (x + 1)(2x - 3)^4$

Core 3

5. Find the equation of the tangent to the curve $y = x\sqrt{1 + 2x}$ at the point $(4, 12)$.

Core 3

6. Use the quotient rule to find $\dfrac{dy}{dx}$, simplifying your answers where necessary.

Part (a) Core 4

   (a) $y = \dfrac{\sin 2x}{x}$ (b) $y = \dfrac{2x - 1}{3x + 4}$ (c) $y = \dfrac{\ln(x + 2)}{x + 2}$

Core 4

7. A curve is defined parametrically by $x = t^2$, $y = t^3$.
   (a) Find $\dfrac{dy}{dx}$ when $t = 2$.
   (b) Find the Cartesian equation of the curve.

Core 4

8. A curve has equation $x^2 + xy^2 - 5 = 0$.
   (a) Find the value of $\dfrac{dy}{dx}$ at $(1, 2)$.
   (b) Find the equation of the normal when $x = 1$, in the form $ax + by + c = 0$.

---

1. (a) $12 \cos 6x$ (b) $-12 \sin x \cos^2 x$ (c) $6 \sec^2 3x$ (d) $\cos x e^{\sin x}$
   (e) $3x^2 \cos(x^3)$
3. (a) $\dfrac{2x}{x^2 + 3}$ (b) $\dfrac{5}{x - 2}$ (c) $-2 \tan 2x$ (d) $\dfrac{6}{6x - 1}$
4. (a) $x(2 \cos 3x - 3x \sin 3x)$ (b) $e^{2x}(\cos x + 2 \sin x)$ (c) $1 + \ln x$ (d) $5(2x + 1)(2x - 3)^3$
5. $3y = 13x - 16$
6. (a) $\dfrac{2x \cos 2x - \sin 2x}{x^2}$ (b) $\dfrac{11}{(3x + 4)^2}$ (c) $\dfrac{1 - \ln(x + 2)}{(x + 2)^2}$
7. (a) 3   (b) $x^3 = y^2$
8. (a) -1.5 (b) $2x - 3y + 4 = 0$

1.6 Integration

After studying this section you should be able to:

- integrate e^x and simple variations of this function with respect to x
- integrate trigonometrical functions
- recognise an integral that leads to a logarithmic function
- integrate using partial fractions
- integrate using a substitution
- use integration by parts to integrate a product
- use integration to find a volume of revolution
- solve first order differential equations when the variables can be separated
- understand exponential growth and decay

The exponential function e^x

Core 3

The fact that differentiation leaves the exponential function unchanged was stated on page 32. Since integration is the reverse process of differentiation it follows that

$$\int e^x \, dx = e^x + c \quad \text{and} \quad \int ae^x \, dx = a\int e^x \, dx = ae^x + c.$$

KEY POINT

An extension of this result, using the reverse of the chain rule, is

$$\int e^{ax+b} \, dx = \frac{1}{a} e^{ax+b} + c.$$

KEY POINT

Examples

1. $\displaystyle\int 4e^x \, dx = 4e^x + c$

2. $\displaystyle\int \tfrac{1}{2} e^{3x-1} \, dx = \tfrac{1}{6} e^{3x-1} + c$

Trigonometric functions

Core 4

The following integrals are obtained by applying the reverse process of differentiating the three main trigonometric functions (page 34).

$$\int \cos x \, dx = \sin x + c \qquad \int \cos(ax+b) \, dx = \frac{1}{a} \sin(ax+b) + c$$

$$\int \sin x \, dx = -\cos x + c \qquad \int \sin(ax+b) \, dx = -\frac{1}{a} \cos(ax+b) + c$$

$$\int \sec^2 x \, dx = \tan x + c \qquad \int \sec^2(ax+b) \, dx = \frac{1}{a} \tan(ax+b) + c$$

You should learn these and practise using them.

Example

$$\int (4\cos 2x - \sin 2x) \, dx = 2 \sin 2x + \tfrac{1}{2} \cos 2x + c$$

It is a good idea to check your answer by differentiating.

Core 3 and Core 4 (Pure Mathematics)

Example

$$\int_{\frac{\pi}{3}}^{\frac{\pi}{2}} \sec^2(\tfrac{1}{2}x) \, dx = \left[\frac{1}{\frac{1}{2}} \tan(\tfrac{1}{2}x)\right]_{\frac{\pi}{3}}^{\frac{\pi}{2}}$$

$$= 2\left[\tan(\tfrac{1}{2}x)\right]_{\frac{\pi}{3}}^{\frac{\pi}{2}}$$

$$= 2\left(\tan\frac{\pi}{4} - \tan\frac{\pi}{6}\right)$$

$$= 2\left(1 - \frac{1}{\sqrt{3}}\right)$$

> *When integrating, the angle must be in radians.*

> *Learn the exact values of the trigonometric ratios of $\frac{\pi}{6}, \frac{\pi}{4}, \frac{\pi}{3}$ (30°, 45°, 60°).*

These results are very useful and can be recognised easily:

$$\frac{d}{dx}(\sin^{n+1} x) = (n+1)\sin^n x \cos x \implies \int \sin^n x \cos x \, dx = \frac{1}{n+1}\sin^{n+1} x + c$$

$$\frac{d}{dx}(\cos^{n+1} x) = -(n+1)\cos^n x \sin x \implies \int \cos^n x \sin x \, dx = -\frac{1}{n+1}\cos^{n+1} x + c$$

Examples

$$\int \sin^5 x \cos x \, dx = \tfrac{1}{6} \sin^6 x + c$$

$$\int \sin x \cos^3 x \, dx = -\tfrac{1}{4} \cos^4 x + c$$

Odd powers of sin x and cos x

Example

Evaluate $\int_0^{\frac{\pi}{6}} \cos^3 x \, dx$

$$\int_0^{\frac{\pi}{6}} \cos^3 x \, dx = \int_0^{\frac{\pi}{6}} \cos x (\cos^2 x) \, dx$$

$$= \int_0^{\frac{\pi}{6}} \cos x (1 - \sin^2 x) \, dx$$

$$= \int_0^{\frac{\pi}{6}} (\cos x - \cos x \sin^2 x) \, dx$$

$$= \left[\sin x - \tfrac{1}{3} \sin^3 x\right]_0^{\frac{\pi}{6}}$$

$$= \tfrac{1}{2} - \tfrac{1}{24} = \tfrac{11}{24}$$

> *Use $\cos^2 x = 1 - \sin^2 x$.*

> *Recognise the integral of $\cos x \sin^2 x$.*

Even powers of cos x and sin x

To integrate even powers of sin x and cos x, use the double angle identities for cos $2x$.

$$\cos 2x = 2\cos^2 x - 1 \implies \cos^2 x = \tfrac{1}{2}(1 + \cos 2x)$$

$$\cos 2x = 1 - 2\sin^2 x \implies \sin^2 x = \tfrac{1}{2}(1 - \cos 2x)$$

Core 3 and Core 4 (Pure Mathematics)

Example

$$\int \cos^2 x \, dx = \tfrac{1}{2} \int (1 + \cos 2x) dx$$
$$= \tfrac{1}{2}(x + \tfrac{1}{2} \sin 2x) + c$$

To integrate $\cos^4 x$, write it as $(\cos^2 x)^2 = (\tfrac{1}{2}(1 + \cos 2x))^2$.

The reciprocal function $\tfrac{1}{x}$

Core 3

The result $\int x^n \, dx = \dfrac{x^{n+1}}{n+1} + c$ holds for any value of n other than $n = -1$.

In this special case, the result takes a different form:

> Remember that $x^{-1} = \dfrac{1}{x}$.

KEY POINT
$$\int \frac{1}{x} \, dx = \ln|x| + c.$$

This extends to $\int \dfrac{k}{x} \, dx = k \int \dfrac{1}{x} \, dx = k \ln|x| + c,$

and to $\int \dfrac{1}{kx} \, dx = \dfrac{1}{k} \int \dfrac{1}{x} \, dx = \dfrac{1}{k} \ln|x| + c.$ (where k is a constant)

For example, $\int \dfrac{2}{x} \, dx = 2 \ln|x| + c$ and $\int \dfrac{1}{3x} \, dx = \dfrac{1}{3} \ln|x| + c.$

Integrals leading to a logarithmic function

Core 4

Remember that $\dfrac{d}{dx} \ln(f(x)) = \dfrac{f'(x)}{f(x)}.$

Applying the process in reverse gives:

KEY POINT
$$\int \frac{f'(x)}{f(x)} \, dx = \ln|f(x)| + c.$$

Examples

(a) $\int \dfrac{x^2}{2x^3 - 2} \, dx = \tfrac{1}{6} \ln|2x^3 - 2| + c$

> $\dfrac{d}{dx}(2x^3 - 2) = 6x^2$ so a factor of $\tfrac{1}{6}$ is needed.

(b) $\int \tan x \, dx = \int \dfrac{\sin x}{\cos x} \, dx$
$= -\ln|\cos x| + c$
$= \ln|\sec x| + c$

> $-\ln a = \ln(a^{-1})$.

Care must be taken with **definite integrals** of this type.
These can be evaluated between the limits a and b only if $f(x)$ exists for all values between a and b. Note that $f(a)$ and $f(b)$ will have the same sign.

$$\int_a^b \frac{f'(x)}{f(x)} \, dx = \Big[\ln|f(x)|\Big]_a^b = (\ln|f(b)| - \ln|f(a)|)$$

Core 3 and Core 4 (Pure Mathematics)

Key points from AS
- **Area under curve**
  Revise AS page 50

Example

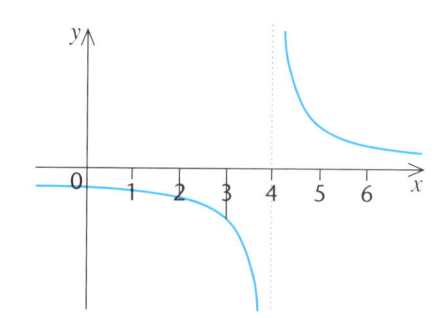

The diagram shows the curve $y = \dfrac{1}{x-4}$.

Find the area between the curve, the x-axis, and the lines $x = 2$ and $x = 3$.

$$\int_a^b y\,dx = \int_2^3 \frac{1}{x-4}\,dx$$
$$= \Big[\ln|x-4|\Big]_2^3$$
$$= \ln|-1| - \ln|-2|$$
$$= \ln 1 - \ln 2$$
$$= -\ln 2\ (= -0.69\ldots)$$

$\ln|-2| = \ln 2$
$\ln 1 = 0$.

Note that the formula gives a negative value, confirming that the area is below the x-axis.

Shaded area = $\ln 2\ (= 0.69\ldots)$ square units.

$f(3)$ and $f(5)$ do not have the same sign.

Note that $\displaystyle\int_3^5 \frac{1}{x-4}\,dx$ cannot be evaluated, as the curve is undefined at $x = 4$.

Partial fractions

Core 4

Some rational functions can be integrated by expressing them in partial fraction form.

Example

Find $\displaystyle\int \frac{x+7}{(x-2)(x+1)}\,dx$, simplifying your answer.

Decompose into partial fractions (see page 15).

$$\frac{x+7}{(x-2)(x+1)} \equiv \frac{3}{x-2} - \frac{2}{x+1}$$

$$\Rightarrow \int \frac{x+7}{(x-2)(x+1)}\,dx = \int\left(\frac{3}{x-2} - \frac{2}{x+1}\right)dx$$

$$= 3\ln(x-2) - 2\ln(x+1) + c$$

$$= \ln k\,\frac{(x-2)^3}{(x+1)^2}$$

The working for this example is shown on page 15.

Key points from AS
- **Logarithms**
  Revise AS page 36

$\log a - \log b = \log\left(\dfrac{a}{b}\right)$
$\log a^n = n\log a$.

Note: Sometimes the constant c is incorporated into the log function by letting $c = \ln k$.

Core 3 and Core 4 (Pure Mathematics)

Substitution

Core 4

Sometimes an integral is made easier by using a substitution.

This transforms the integral with respect to one variable, say x, into an integral with respect to a related variable, say u.

$$\int f(x)\,dx = \int f(x)\frac{dx}{du}\,du.$$

KEY POINT

This method is known as the method of **substitution** or **change of variable**.

Example

Find $\int x\sqrt{2x+1}\,dx$, using the substitution $u = \sqrt{2x+1}$.

This could also be done by using the substitution $u = 2x + 1$.

$$\int x\sqrt{2x+1}\,dx = \int x\sqrt{2x+1}\,\frac{dx}{du}\,du$$

Side working

$$= \int \tfrac{1}{2}(u^2 - 1)u \times u\,du \qquad u = \sqrt{2x+1}$$

Show the side working as part of your answer.

$$= \frac{1}{2}\int (u^4 - u^2)\,du \qquad u^2 = 2x+1 \Rightarrow x = \tfrac{1}{2}(u^2 - 1)$$

$$= \frac{1}{2}\left(\frac{u^5}{5} - \frac{u^3}{3}\right) + c \qquad \frac{dx}{du} = u$$

When simplifying, take out factors as soon as possible.

$$= \frac{u^3}{2}\left(\frac{u^2}{5} - \frac{1}{3}\right) + c$$

$$= \tfrac{1}{30}u^3(3u^2 - 5) + c \qquad 3u^2 - 5 = 3(2x+1) - 5$$

Write the answer in terms of x, simplifying as much as possible.

$$= \tfrac{1}{30}(2x+1)^{\frac{3}{2}}(6x - 2) + c \qquad = 6x - 2$$

$$= \tfrac{1}{15}(2x+1)^{\frac{3}{2}}(3x - 1) + c \qquad = 2(3x - 1)$$

When evaluating definite integrals using the method of substitution, change the x limits to u limits and substitute them into the working as soon as possible. There is no need to tidy up the expression first.

Sometimes it is possible to bypass the method of substitution by **recognising** an integral as the derivative of a particular function.

Examples

This type of integral can always be done by using a substitution.

(a) $\int (2x+1)^5\,dx = \tfrac{1}{12}(2x+1)^6 + c$

Recognise that
$\frac{d}{dx}(2x+1)^6 = 12(2x+1)^5$.

(b) $\int x\sqrt{2x^2+1}\,dx = \tfrac{1}{6}(2x^2+1)^{\frac{3}{2}} + c$

Recognise that
$\frac{d}{dx}(2x^2+1)^{\frac{3}{2}} = 6x(2x^2+1)^{\frac{1}{2}}$.

General result:

$$\int f'(x)[f(x)]^n\,dx = \frac{1}{n+1}[f(x)]^{n+1} + c$$

Core 3 and Core 4 (Pure Mathematics)

Integration by parts

Core 4

It is sometimes possible to integrate a product using **integration by parts**.

> **KEY POINT**
> If u and v are functions of x, then $\int u \dfrac{dv}{dx} dx = uv - \int v \dfrac{du}{dx} dx$.

Before using integration by parts, check whether the function can be simplified, or whether the integral can be recognised directly or done using a substitution.

Example

Find $\int 2x \sin x \, dx$.

Let $u = 2x$, then $\dfrac{du}{dx} = 2$

Let $\dfrac{dv}{dx} = \sin x$ then $v = -\cos x$

> *Integration by parts can only be used if v can be found from $\dfrac{dv}{dx}$ and it is possible to find $\int v \dfrac{du}{dx} dx$.*

$\therefore \int 2x \sin x \, dx = 2x(-\cos x) - \int (-\cos x \times 2) dx$

$= -2x \cos x + \int 2 \cos x \, dx$

$= -2x \cos x + 2 \sin x + c$

Sometimes the integration by parts process has to be carried out more than once. This happens when integrating expressions such as $x^2 e^x$ or $x^2 \cos x$.

Definite integrals

$$\int_a^b u \dfrac{dv}{dx} dx = \left[uv \right]_a^b - \int_a^b v \dfrac{du}{dx} dx$$

Example

$\int_0^1 x e^{2x} dx = [x \tfrac{1}{2} e^{2x}]_0^1 - \int_0^1 \tfrac{1}{2} e^{2x} dx$

$= \tfrac{1}{2} e^2 - [\tfrac{1}{4} e^{2x}]_0^1$

$= \tfrac{1}{2} e^2 - (\tfrac{1}{4} e^2 - \tfrac{1}{4}) = \tfrac{1}{4} e^2 + \tfrac{1}{4}$

Working:

Let $u = x$, then $\dfrac{du}{dx} = 1$

Let $\dfrac{dv}{dx} = e^{2x}$, then $v = \tfrac{1}{2} e^{2x}$

Some functions containing $\ln x$ can be integrated using integration by parts. A useful strategy is to take $\ln x$ as u.

> *Often a polynomial function is taken as u because it becomes of lower degree when differentiated. Integrals involving $\ln x$ are an exception, since $\ln x$ is not easy to integrate, but can be differentiated.*

Example

$\int x^2 \ln x \, dx = \int (\ln x \times x^2) dx$

$= \ln x \times \left(\dfrac{x^3}{3}\right) - \int \left(\dfrac{x^3}{3} \times \dfrac{1}{x}\right) dx$

$= \tfrac{1}{3} x^3 \ln x - \tfrac{1}{3} \int x^2 \, dx$

$= \tfrac{1}{3} x^3 \ln x - \tfrac{1}{9} x^3 + c$

Working:

Let $u = \ln x$, then $\dfrac{du}{dx} = \dfrac{1}{x}$

Let $\dfrac{dv}{dx} = x^2$, then $v = \dfrac{x^3}{3}$

Core 3 and Core 4 (Pure Mathematics)

This method is particularly useful when integrating $\ln x$.
Think of it as integrating the product $\ln x \times 1$ as follows:

$$\int (\ln x \times 1) dx = \ln x \times x - \int \left(x \times \frac{1}{x}\right) dx$$

$$= x \ln x - \int 1 dx$$

$$= x \ln x - x + c$$

Working:

Let $u = \ln x$, then $\dfrac{du}{dx} = \dfrac{1}{x}$

Let $\dfrac{dv}{dx} = 1$, then $v = x$

Volume of revolution

Core 3

The shaded region R is bounded by the curve $y = f(x)$, the x-axis and the lines $x = a$ and $x = b$.

Rotating R completely about the x-axis forms a solid figure. The volume of this figure is called the volume of revolution and is given by

$$V = \int_a^b \pi y^2 \, dx$$

The corresponding result for rotation about the y-axis is

$$\int_a^b \pi x^2 \, dy$$

where the region is bounded by the curve $y = f(x)$, the y-axis and the lines $y = a$ and $y = b$.

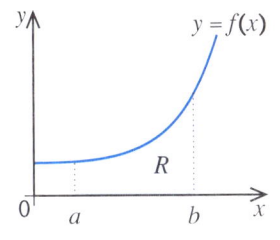

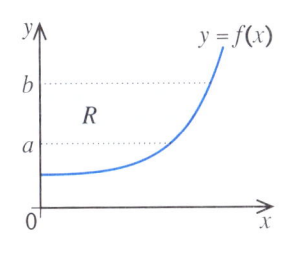

Example Find a formula for the volume of a cone of height h and base radius r.

The volume of the cone is given by the volume of revolution of the shaded region shown.

The straight line has gradient $\dfrac{r}{h}$ and passes through the origin, so its equation is $y = \dfrac{r}{h} x$.

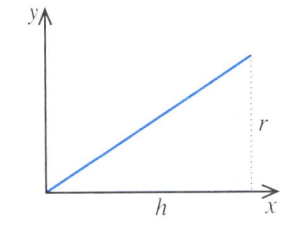

Using the formula for volume of revolution gives:

$$V = \int_0^h \pi \left(\frac{r}{h} x\right)^2 dx = \frac{\pi r^2}{h^2} \int_0^h x^2 \, dx$$

$$= \frac{\pi r^2}{h^2} \left[\frac{x^3}{3}\right]_0^h$$

$$= \frac{\pi r^2}{h^2} \times \frac{h^3}{3}$$

$$= \frac{\pi r^2 h}{3}.$$

Core 3 and Core 4 (Pure Mathematics)

Differential equations

Core 4

A **first order differential equation** in x and y contains only the first differential coefficient.

Examples $\quad \dfrac{dy}{dx} = -2x, \quad \dfrac{dy}{dx} = \dfrac{x}{y}, \quad y\dfrac{dy}{dx} = 3$

If $\dfrac{dy}{dx} = -2x$, then integrating with respect to x gives $y = -x^2 + c$.

> The general solution contains an arbitrary integration constant.

This is the **general solution** of the differential equation and it can be illustrated by a family of curves.

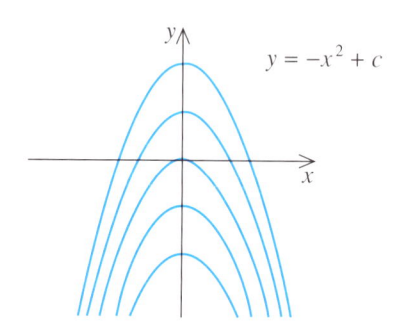

$y = -x^2 + c$

A **particular solution** is a specific member of the family and it can be found from additional information.

> The value of c is found to give a particular solution.

For example, if the curve $y = -x^2 + c$ goes through the point $(2, 0)$, then $0 = -(2)^2 + c \Rightarrow c = 4$. The particular solution is $y = -x^2 + 4$.

Separating the variables

> **KEY POINT**
> A differential equation that can be written in the form $f(y)\dfrac{dy}{dx} = g(x)$ can be solved by **separating the variables**, where $\displaystyle\int f(y)dy = \int g(x)dx$.

Example

Solve (a) $\dfrac{dy}{dx} = \dfrac{x}{y}$ (b) $\dfrac{dy}{dx} = y$

(a) $\dfrac{dy}{dx} = \dfrac{x}{y}$

> Separate the variables to get the format $\int \ldots dy = \int \ldots dx$.

$\displaystyle\int y \, dy = \int x \, dx \Rightarrow \tfrac{1}{2}y^2 = \tfrac{1}{2}x^2 + c$

(b) $\dfrac{dy}{dx} = y$

> This is a special case, when $g(x) = 1$.

$\dfrac{1}{y}\dfrac{dy}{dx} = 1$

$\displaystyle\int \dfrac{1}{y} dy = \int 1 dx$

> The answer can be left like this.

$\ln y = x + c$

$y = e^{x+c} = e^x \times e^c = Ae^x \quad$ (where $A = e^c$)

> The letter A is often used here, but it could be any letter (other than e, x or y, of course!).

So $\dfrac{dy}{dx} = y \Rightarrow y = Ae^x$

Core 3 and Core 4 (Pure Mathematics)

Exponential growth and decay

Core 4

You may have to form a differential equation from given information. Often this models exponential growth or exponential decay.

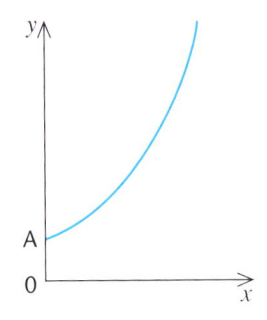

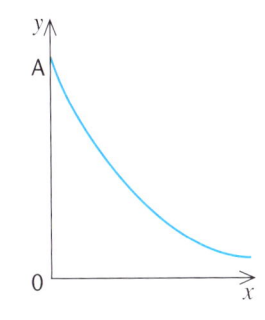

Exponential growth

Exponential decay

In exponential decay, the fact that y is decreasing with respect to x is usually shown by the negative, keeping $k > 0$.

$$\frac{dy}{dx} = ky \Rightarrow y = Ae^{kx} \quad (k > 0)$$

$$\frac{dy}{dx} = -ky \Rightarrow y = Ae^{-kx} \quad (k > 0)$$

An example of exponential growth is the growth of a population where

p_0 is the initial population.

$$\frac{dp}{dt} = kt \Rightarrow p = p_0 e^{-kt}$$

Examples of exponential decay are:

m_0 is the initial mass.

(a) the disintegration of radioactive materials, $\dfrac{dm}{dt} = -km \Rightarrow m = m_0 e^{-kt}$

T_s is temperature of surroundings, T_0 is initial temperature of the object.

(b) Newton's Law of Cooling, $\dfrac{dT}{dt} = -k(T - T_s) \Rightarrow T - T_s = (T_0 - T_s)e^{-kt}$

Core 3 and Core 4 (Pure Mathematics)

Progress check

Core 3

1. Integrate with respect to x.

   (a) $\dfrac{3}{x}$ (b) $\dfrac{x+3}{x^2}$ (c) $3e^x - \dfrac{1}{x}$ (d) $3e^{4-x}$

Core 3

2. The region bounded by the curve $y = e^{\frac{x}{2}}$, the lines $x = 1$ and $x = 3$ and the x-axis is rotated completely about the x-axis. Find the volume of revolution.

Core 4

3. (a) $\int 4 \sin 2x \, dx$ (b) $\int \sec^2 2x \, dx$ (c) $\int \cos(\tfrac{1}{2}x) \, dx$

   (d) $\int 4 \sin^2 x \, dx$ (e) $\int \cos^2 2x \, dx$ (f) Evaluate $\int_0^{\frac{1}{6}\pi} \sin^4 x \cos x \, dx$

Part (a) Core 3

Core 4

4. Find (a) $\int \dfrac{1}{2x+5} \, dx$

   (b) $\int \cot x \, dx$

   (c) $\int \dfrac{3x+2}{x^2 - x - 12} \, dx$ *Split (c) into partial fractions first.*

Core 4

5. Using the substitution $u = 2 + e^{2x}$, or otherwise, find $\int \dfrac{e^{2x}}{2 + e^{2x}} \, dx$ *Remember that $\dfrac{dx}{du} = \dfrac{1}{du/dx}$.*

Core 4

6. (a) Find (i) $\int xe^{-x} \, dx$ (ii) $\int x^3 \ln x \, dx$ (b) Evaluate $\int_0^{\frac{1}{2}\pi} x \cos x \, dx$

Core 4

7. Solve the differential equation $\dfrac{dy}{dx} = 2xy$, given that $y = 3$ when $x = 0$.

---

1. (a) $3 \ln|x| + c$ (b) $\ln|x| - \dfrac{3}{x} + c$ (c) $3e^x - \ln|x| + c$ (d) $-3e^{4-x} + c$
2. $\int_1^3 \pi e^x \, dx = \pi[e^x]_1^3 = \pi(e^3 - e)$.
3. (a) $-2 \cos 2x + c$ (b) $\tfrac{1}{2} \tan 2x + c$ (c) $2 \sin(\tfrac{1}{2}x) + c$
   (d) $2x - \sin 2x + c$ (e) $\tfrac{1}{2}x + \tfrac{1}{8} \sin 4x + c$ (f) $\dfrac{1}{160}$
4. (a) $\tfrac{1}{2} \ln(2x+5) + c$ (b) $\ln(\sin x) + c$ (c) $\ln(x+3)(x-4)^2 + c$ or $\ln A(x+3)(x-4)^2$
5. $\tfrac{1}{2} \ln(2 + e^{2x}) + c$
6. (a) (i) $-e^{-x}(x+1) + c$ (ii) $\tfrac{1}{16}x^4(4 \ln x - 1) + c$ (b) $\tfrac{1}{2}\pi - 1$
7. $y = 3e^{x^2}$

Core 3 and Core 4 (Pure Mathematics)

1.7 Numerical methods

After studying this section you should be able to:
- locate an interval containing a root of an equation by finding a change of sign
- use simple iteration to solve equations
- find the approximate value of an integral using Simpson's rule

LEARNING SUMMARY

Change of sign

Core 3

If there is an interval from $x = a$ to $x = b$ in which the graph of a function has no breaks then the function is said to be **continuous** on the interval.

All polynomial functions are continuous everywhere.

For example, the graph of $y = x^2 - 3$ is continuous everywhere.

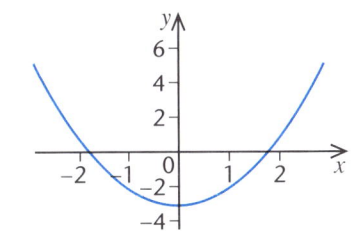

However, the graph of $y = \dfrac{1}{x-3}$ is not continuous on any interval that contains the value $x = 3$.

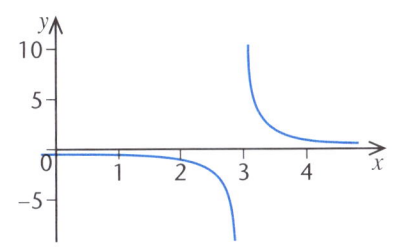

> If $f(x)$ is continuous between $x = a$ and $x = b$ and if $f(a)$ and $f(b)$ have different signs, then a root of $f(x) = 0$ lies in the interval from a to b.

KEY POINT

Example Given that $f(x) = e^x - 10x$, show that the equation $f(x) = 0$ has a root between 3 and 4.

$f(3) = -9.914\ldots < 0$
$f(4) = 14.598\ldots > 0$.

The change of sign shows that $f(x) = 0$ has a root between 3 and 4.

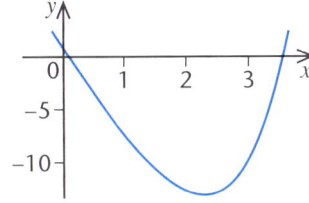

You can continue this process, called a **decimal search**, to trap the root in a smaller and smaller interval. The table gives the values of the function in steps of 0.1.

Most graphic calculators can tabulate values of a function and this gives a very efficient way to obtain the information.

x	3.1	3.2	3.3	3.4	3.5	3.6	3.7	3.8	3.9
$f(x)$	−8.802	−7.467	−5.887	−4.035	−1.884	0.5982	3.4473	6.7011	10.402

This shows that the root lies between 3.5 and 3.6
At the mid-point $f(3.55) = -0.6866\ldots < 0$ and so the root lies between 3.55 and 3.6

This is a good way to establish the value of a root to a particular level of accuracy.

3.5   3.55   3.6

It follows that the value of the root is 3.6 to 1 d.p.

49

Core 3 and Core 4 (Pure Mathematics)

Iteration

Core 3

To solve an equation by **iteration** you start with some approximation to a root and improve its accuracy by substituting it into a formula. You can then repeat the process until you have the desired level of accuracy.

> There are different methods for producing an iterative formula. You only need to know about simple iteration for this component.

Using **simple iteration**, the first step is to rearrange the equation to express x as a function of itself. This function defines the iterative formula that you need.

For example, the equation $e^x - 10x = 0$ can be rearranged as

$$e^x = 10x$$
$$\Rightarrow x = \ln(10x).$$

This may now be turned into an iterative formula for finding x.

$$x_{n+1} = \ln(10x_n).$$

Starting with $x_1 = 3$, $x_2 = \ln 30 = 3.401\ldots$ and so on. This produces a sequence of values:

> The values may be found very easily using a calculator's Ans function: Key in 3 = followed by ln(10Ans) = = = to produce the sequence. (Depending on the make of your calculator you may need to use EXE or ENTER in place of = .)

$x_2 = 3.401\ldots$ $x_6 = 3.5760\ldots$
$x_3 = 3.526\ldots$ $x_7 = 3.5768\ldots$
$x_4 = 3.563\ldots$ $x_8 = 3.5770\ldots$
$x_5 = 3.573\ldots$ $x_9 = 3.5771\ldots$

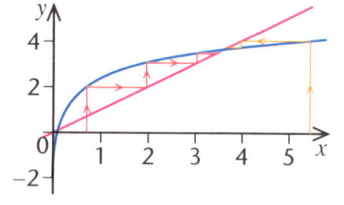

The calculator display soon settles on 3.577152064. The value of the root is $x = 3.577$ to 4 s.f.

The diagram shows how the process converges from the starting point on either side of the root.

> Start from a point on the x-axis, move up to the curve and across to $y = x$. Then move up to the curve and across to $y = x$ again. Continue in the same way. Each of these stages corresponds to one iteration of the formula.

A different rearrangement of the original equation gives $x = \dfrac{e^x}{10}$ and so the corresponding iterative formula is $x_{n+1} = \dfrac{e^{x_n}}{10}$.

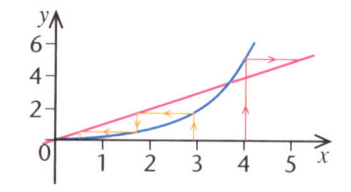

> You may need to try several rearrangements in order to converge to a particular root of an equation by this method.

Starting with $x = 3$, this rearrangement fails to converge to the root between 3 and 4 but it does converge to the other root of the equation.

> Notice how the movement is away from the upper root this time. If the starting value is smaller than the upper root then the process converges to the lower root. A starting value above the upper root fails to converge to either root.

The value of this root is $x = 0.1118$ to 4 s.f.

Example Starting with $x_1 = 1$, use the formula $x_{n+1} = \dfrac{x_n^2 - 5x_n}{10} - 3$ to find $x_2, x_3, \ldots, x_8$ and describe the long term behaviour of the sequence.

> Try the key sequence:
> 1 =
> (Ans² − 5Ans)/10 − 3
> =
> =
> =
> =

$x_2 = -3.4$
$x_3 = -0.144$
$x_4 = -2.9259264$
$x_5 = -0.6809322$
$x_6 = -2.6131669$
$x_7 = -1.0105523$
$x_8 = -2.3926032$

The sequence oscillates but converges to −1.787 to 4 s.f. which is the lower root of the equation

$$x = \frac{x^3 - 5x}{10} - 3,$$

which simplifies to

$$x^2 - 15x - 30 = 0.$$

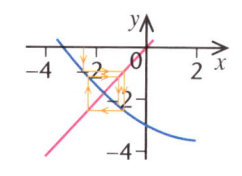

The oscillations correspond to a cobweb pattern on the diagram.

Core 3 and Core 4 (Pure Mathematics)

Numerical integration

Core 3

Simpson's rule

This method divides the area into an **even number** of parallel strips and approximates the areas of pairs of strips using the following formula:

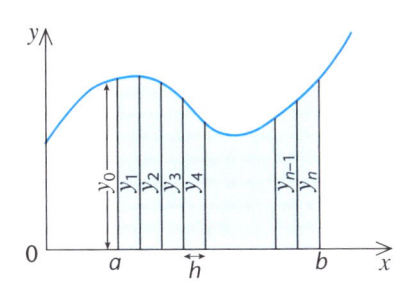

$$\int_a^b f(x)\,dx \approx \tfrac{1}{3} h\{(y_0 + y_n) + 4(y_1 + y_3 + \ldots + y_{n-1}) + 2(y_2 + y_4 + \ldots + y_{n-2})\}$$

where $h = \dfrac{b-a}{n}$ and n is even.

KEY POINT

Note that the first ordinate has been labelled y_0 and the last y_n. There are n strips and $n+1$ ordinates.

Example

Estimate $\displaystyle\int_0^1 e^{x^2}\,dx$ using Simpson's rule with 10 strips.

Tabulating the results helps in doing the final calculation.

x	y	First and last ordinates	Odd ordinates	Other ordinates
0	y_0	1		
0.1	y_1		1.010 ...	
0.2	y_2			1.040 ...
0.3	y_3		1.094 ...	
0.4	y_4			1.173 ...
0.5	y_5		1.284 ...	
0.6	y_6			1.433 ...
0.7	y_7		1.632 ...	
0.8	y_8			1.896 ...
0.9	y_9		2.247 ...	
1	y_{10}	2.718 ...		
Totals		3.718 ...	7.268 ...	5.544 ...

If you have multiple memories in your calculator, retain the figures. Otherwise approximate say to 4 decimal places.

$h = \dfrac{1-0}{10} = 0.1$

$\displaystyle\int_a^b f(x)\,dx \approx \tfrac{1}{3} h\{(y_0 + y_{10}) + 4(y_1 + y_3 + y_5 + y_7 + y_9) + 2(y_2 + y_4 + y_6 + y_8)\}$

$= \tfrac{1}{3} \times 0.1 \times \{3.718\ldots + 4(7.268\ldots) + 2(5.544\ldots)\}$

$= 1.463$ (3 s.f.)

Core 3 and Core 4 (Pure Mathematics)

Progress check

Core 3

1. Show that the equation $x^3 - x - 7 = 0$ has a root between 2 and 3. Use a decimal search to find the value of the root to 2 d.p.

Core 3

2. Use simple iteration to refine the value of the root found in question 1 and state its value to 4 d.p.

Core 3

3. Estimate $\int_0^1 \sqrt{1-x^3}\,dx$ using Simpson's rule with 4 strips.

1. 2.09
2. 2.0867
3. To 3 d.p. 0.823

Core 3 and Core 4 (Pure Mathematics)

1.8 Vectors

LEARNING SUMMARY

After studying this section you should be able to:
- understand the distinction between vector and scalar quantities
- add and subtract vector quantities and multiply by a scalar
- use the unit vectors **i**, **j** and **k**
- find the magnitude and direction of a vector
- use scalar products to find the angle between two directions
- express equations of lines in vector form
- determine whether two lines are parallel, intersect or are skew

Vector and scalar quantities

Core 4

A **scalar** quantity has size (or **magnitude**) but not direction. **Numbers** are scalars and some other important examples are **distance**, **speed**, **mass** and **time**.

A **vector** quantity has both size and **direction**. For example, distance in a specified direction is called **displacement**.

The diagram shows a **directed line segment**. It has size (in this case, length) and direction so it is a vector.

The diagram gives a useful way to represent *any* vector quantity and may also be used to represent addition and subtraction of vectors and multiplication of a vector by a scalar.

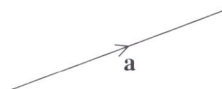

> In a textbook, vectors are usually labelled with lower case letters in **bold** print. When hand-written, these letters should be underlined e.g. $\underline{a}$.

Addition and subtraction of vectors

Core 4

This diagram shows three vectors **a**, **b** and **c** such that **c** = **a** + **b**.

The vectors **a** and **b** follow on from each other and then **c** joins the start of vector **a** to the end of vector **b**.

Check that **a** + **b** = **b** + **a**.

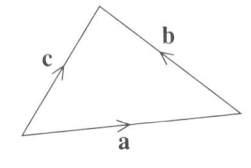

The vector **c** is called the **resultant** of **a** and **b**.

On a vector diagram, −**q** has the opposite sense of direction to **q**.
Notice that **p** − **q** is represented as **p** + (−**q**).

This diagram shows the same information in a different way.

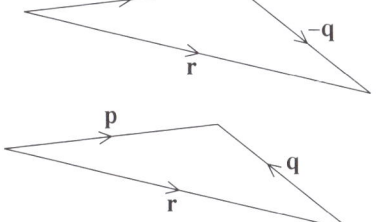

Following the route in the opposite direction to **q** is the same as adding −**q** or subtracting **q**.

53

Core 3 and Core 4 (Pure Mathematics)

Scalar multiplication

Core 4

2**p** is parallel to **p** and has the same sense of direction but is twice as long.

−3**p** is parallel to **p** but has the opposite sense of direction and is three times as long.

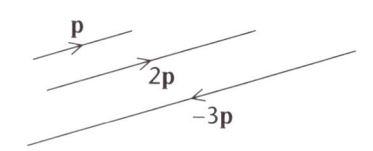

Component form

Core 4

When working in two dimensions, it is often very useful to express a vector in terms of two special vectors **i** and **j**. These are **unit vectors** at right-angles to each other. A vector **r** written as **r** = a**i** + b**j** is said to have **components** a**i** and b**j**.

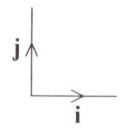

> A unit vector is a vector of magnitude 1 unit.

Examples

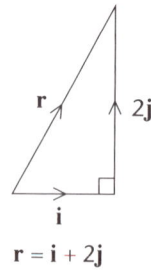

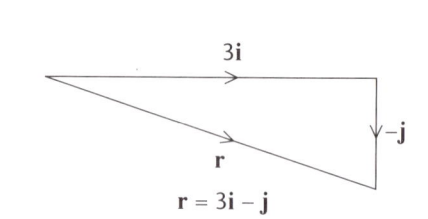

r = **i** + 2**j**

r = 3**i** − **j**

For work involving the Cartesian coordinate system, **i** and **j** are taken to be in the positive directions of the x- and y-axes respectively.

In three dimensions, a third vector **k** is used to represent a unit vector in the positive direction of the z-axis.

The **position vector** of a point P is the vector $\overrightarrow{OP}$ where O is the origin.

If P has coordinates (a, b) then its position vector is given by **r** = a**i** + b**j**.

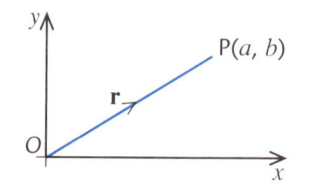

In three dimensions, a point with coordinates (a, b, c) would have position vector **r** = a**i** + b**j** + c**k**.

Adding and subtracting vectors in component form

Core 4

One advantage of expressing vectors in terms of **i**, **j** and **k** is that addition and subtraction may be done algebraically.

Example

p = 3**i** + 2**j** − **k** and **q** = 5**i** − **j** + 4**k**. Express the following vectors in terms of **i**, **j** and **k**:

(a) **p** + **q**   (b) **p** − **q**   (c) 2**p** − 3**q**.

(a) **p** + **q** = (3**i** + 2**j** − **k**) + (5**i** − **j** + 4**k**) = 8**i** + **j** + 3**k**

(b) **p** − **q** = (3**i** + 2**j** − **k**) − (5**i** − **j** + 4**k**) = −2**i** + 3**j** − 5**k**

(c) 2**p** − 3**q** = 2(3**i** + 2**j** − **k**) − 3(5**i** − **j** + 4**k**)
  = 6**i** + 4**j** − 2**k** − 15**i** + 3**j** − 12**k** = −9**i** + 7**j** − 14**k**

Core 3 and Core 4 (Pure Mathematics)

Finding the magnitude of a vector

Core 4

> **KEY POINT**
>
> If $\mathbf{r} = a\mathbf{i} + b\mathbf{j} + c\mathbf{k}$, the magnitude (length) of $\mathbf{r}$ is given by
> $$|\mathbf{r}| = \sqrt{a^2 + b^2 + c^2}$$

For example,
If $\mathbf{r} = 3\mathbf{i} - 4\mathbf{j} + 12\mathbf{k}$, then $|\mathbf{r}| = \sqrt{3^2 + (-4)^2 + 12^2} = 13$.

Scalar product

Core 4

If $\mathbf{a} = x_1\mathbf{i} + y_1\mathbf{j} + z_1\mathbf{k}$, then $|\mathbf{a}| = \sqrt{x_1^2 + y_1^2 + z_1^2}$ where $|\mathbf{a}|$ is the length of $\mathbf{a}$.
If $\mathbf{b} = x_2\mathbf{i} + y_2\mathbf{j} + z_2\mathbf{k}$, then $|\mathbf{b}| = \sqrt{x_2^2 + y_2^2 + z_2^2}$.

> **KEY POINT**
>
> If θ is the angle between $\mathbf{a}$ and $\mathbf{b}$,
> the **scalar** (or **dot**) **product** of $\mathbf{a}$ and $\mathbf{b}$ is $\mathbf{a}.\mathbf{b}$
> where $\mathbf{a}.\mathbf{b} = |\mathbf{a}||\mathbf{b}|\cos\theta$.

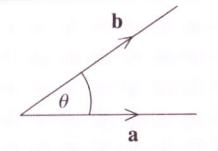

This leads to the useful results:
$\mathbf{i}.\mathbf{i} = 1$, $\mathbf{j}.\mathbf{j} = 1$, $\mathbf{k}.\mathbf{k} = 1$
$\mathbf{i}.\mathbf{j} = \mathbf{j}.\mathbf{k} = \mathbf{k}.\mathbf{i} = 0$.

If two vectors are **parallel**, $\theta = 0$ and $\mathbf{a}.\mathbf{b} = |\mathbf{a}||\mathbf{b}|$.
If two vectors are **perpendicular**, $\theta = 90°$ and $\mathbf{a}.\mathbf{b} = 0$.

In column vectors
$$\mathbf{a}.\mathbf{b} = \begin{pmatrix} x_1 \\ y_1 \\ z_1 \end{pmatrix} \cdot \begin{pmatrix} x_2 \\ y_2 \\ z_2 \end{pmatrix}$$
$= x_1 x_2 + y_1 y_2 + z_1 z_2$.

For $\mathbf{a}$ and $\mathbf{b}$ defined as above:

$\mathbf{a}.\mathbf{b} = (x_1\mathbf{i} + y_1\mathbf{j} + z_1\mathbf{k}).(x_2\mathbf{i} + y_2\mathbf{j} + z_2\mathbf{k})$
$\quad = x_1 x_2 + y_1 y_2 + z_1 z_2$

This uses the results for parallel and perpendicular vectors.

Example
Find the angle between $\mathbf{a} = 2\mathbf{i} + \mathbf{j} + 4\mathbf{k}$ and $\mathbf{b} = -3\mathbf{i} + 2\mathbf{j} - \mathbf{k}$.

$\mathbf{a}.\mathbf{b} = 2(-3) + 1(2) + 4(-1) = -8$
$|\mathbf{a}| = \sqrt{2^2 + 1^2 + 4^2} = \sqrt{21}$
$|\mathbf{b}| = \sqrt{(-3)^2 + 2^2 + (-1)^2} = \sqrt{14}$
$\mathbf{a}.\mathbf{b} = |\mathbf{a}||\mathbf{b}|\cos\theta$
$-8 = \sqrt{21}\sqrt{14}\cos\theta$
$\Rightarrow \cos\theta = \dfrac{-8}{\sqrt{21}\sqrt{14}} = -0.4665\ldots$
$\Rightarrow \theta = 117.8°$ (1 d.p.)

In column vectors:
$$\mathbf{a}.\mathbf{b} = \begin{pmatrix} 2 \\ 1 \\ 4 \end{pmatrix} \cdot \begin{pmatrix} -3 \\ 2 \\ -1 \end{pmatrix} = -8.$$

Vector equation of a line

Core 4

> **KEY POINT**
>
> A **vector equation** of a line passing through a fixed point A with position vector $\mathbf{a}$ and parallel to a vector $\mathbf{b}$ is
> $\mathbf{r} = \mathbf{a} + t\mathbf{b}$, where t is a scalar parameter.

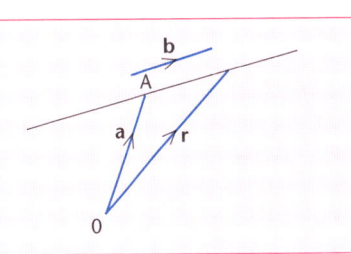

t is often used for the parameter, but not exclusively so. Other letters sometimes used include s, μ and λ.

55

Core 3 and Core 4 (Pure Mathematics)

r is the position vector of any point on the line.
b is the **direction vector** of the line.

The equation of the line is not unique. Any other point on the line could be used instead of A and any multiple of **b** could be used for the direction vector.

In column format:
$$\mathbf{r} = \begin{pmatrix} 4 \\ -5 \\ 1 \end{pmatrix} + t\begin{pmatrix} 3 \\ 4 \\ -2 \end{pmatrix}$$

Example
A vector equation of the line passing through $(4, -5, 1)$, parallel to $3\mathbf{i} + 4\mathbf{j} - 2\mathbf{k}$ is
$\mathbf{r} = 4\mathbf{i} - 5\mathbf{j} + \mathbf{k} + t(3\mathbf{i} + 4\mathbf{j} - 2\mathbf{k})$.

> **KEY POINT**
> A vector equation of the line through two fixed points A and B, with position vectors **a** and **b** is given by
> $\mathbf{r} = \mathbf{a} + t(\mathbf{b} - \mathbf{a})$ where t is a scalar parameter.

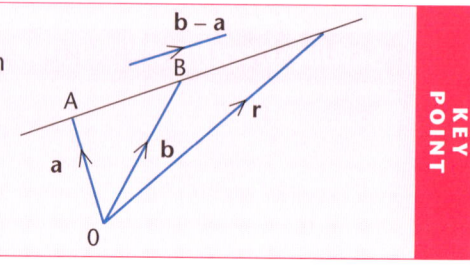

Example
Find a vector equation of the line through the points $P(3, 1, -4)$ and $Q(-2, 5, 1)$.
Direction of line PQ:

$\overrightarrow{PQ} = \mathbf{q} - \mathbf{p} = -2\mathbf{i} + 5\mathbf{j} + \mathbf{k} - (3\mathbf{i} + \mathbf{j} - 4\mathbf{k}) = -5\mathbf{i} + 4\mathbf{j} + 5\mathbf{k}$

A vector equation of PQ is $\mathbf{r} = 3\mathbf{i} + \mathbf{j} - 4\mathbf{k} + t(-5\mathbf{i} + 4\mathbf{j} + 5\mathbf{k})$.

Angles between two lines

To find the angle between two lines, find the angle between their direction vectors.

Pairs of lines

Core 4

In 2-dimensions, a pair of lines are either parallel or they intersect.
In 3-dimensions, a pair of lines are parallel, or they intersect or they are **skew**.

If two lines are parallel, then their direction vectors are multiples of each other, as in this **example**:
$\mathbf{r} = 3\mathbf{i} - 5\mathbf{j} + 2\mathbf{k} + \lambda(\mathbf{i} + 2\mathbf{j} - \mathbf{k})$ and $\mathbf{r} = 4\mathbf{i} + \mathbf{j} - 2\mathbf{k} + \mu(3\mathbf{i} + 6\mathbf{j} - 3\mathbf{k})$.

The two lines $\mathbf{r} = \mathbf{a} + \lambda\mathbf{b}$ and $\mathbf{r} = \mathbf{c} + \mu\mathbf{d}$ intersect if unique values of λ and μ can be found such that $\mathbf{a} + \lambda\mathbf{b} = \mathbf{c} + \mu\mathbf{d}$.
If unique values cannot be found, then the two lines are **skew**.

Example
Investigate whether the lines $\mathbf{r}_1 = 2\mathbf{i} + \mathbf{j} - 3\mathbf{k} + \lambda(4\mathbf{i} + 6\mathbf{j} - \mathbf{k})$ and $\mathbf{r}_2 = 3\mathbf{i} - 2\mathbf{k} + \mu(4\mathbf{i} + \mathbf{j} + 3\mathbf{k})$ intersect or whether they are skew.

$\mathbf{r}_1 = \mathbf{r}_2 \implies 2\mathbf{i} + \mathbf{j} - 3\mathbf{k} + \lambda(4\mathbf{i} + 6\mathbf{j} - \mathbf{k}) = 3\mathbf{i} - 2\mathbf{k} + \mu(4\mathbf{i} + \mathbf{j} + 3\mathbf{k})$

i.e. $(2 + 4\lambda)\mathbf{i} + (1 + 6\lambda)\mathbf{j} + (-3 - \lambda)\mathbf{k} = (3 + 4\mu)\mathbf{i} + (0 + \mu)\mathbf{j} + (-2 + 3\mu)\mathbf{k}$

If the lines intersect, then $\mathbf{r}_1 = \mathbf{r}_2$ will have a unique solution for λ and μ.

Note that the lines are not parallel, since their direction vectors are not multiples of each other.

Equating coefficients:

$2 + 4\lambda = 3 + 4\mu$ (1)
$1 + 6\lambda = \mu$ (2)
$-3 - \lambda = -2 + 3\mu$ (3)

Solving (1) and (2) simultaneously gives:
$\lambda = -0.25$ and $\mu = -0.5$
Check whether these satisfy (3):
LHS $= -3 - (-0.25) = -2.75$
RHS $= -2 + 3(-0.5) = -3.5$

∴ λ = −0.25 and μ = −0.5 do not satisfy all three equations.
Since there is not a unique solution for λ and μ, the lines do not intersect. They are skew.

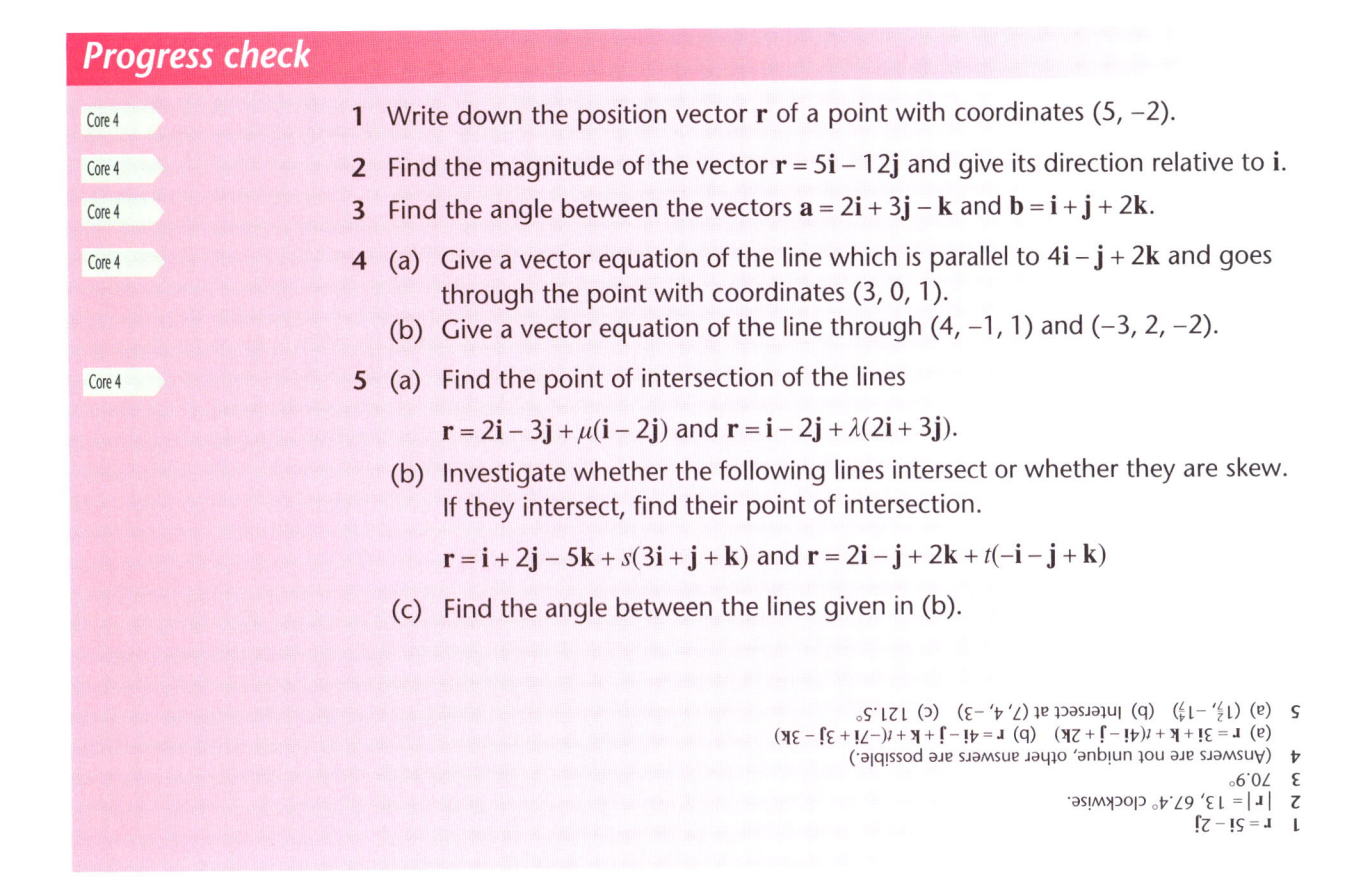

Progress check

Core 4

1 Write down the position vector **r** of a point with coordinates (5, −2).

Core 4

2 Find the magnitude of the vector **r** = 5**i** − 12**j** and give its direction relative to **i**.

Core 4

3 Find the angle between the vectors **a** = 2**i** + 3**j** − **k** and **b** = **i** + **j** + 2**k**.

Core 4

4 (a) Give a vector equation of the line which is parallel to 4**i** − **j** + 2**k** and goes through the point with coordinates (3, 0, 1).
(b) Give a vector equation of the line through (4, −1, 1) and (−3, 2, −2).

Core 4

5 (a) Find the point of intersection of the lines

$$\mathbf{r} = 2\mathbf{i} - 3\mathbf{j} + \mu(\mathbf{i} - 2\mathbf{j}) \text{ and } \mathbf{r} = \mathbf{i} - 2\mathbf{j} + \lambda(2\mathbf{i} + 3\mathbf{j}).$$

(b) Investigate whether the following lines intersect or whether they are skew. If they intersect, find their point of intersection.

$$\mathbf{r} = \mathbf{i} + 2\mathbf{j} - 5\mathbf{k} + s(3\mathbf{i} + \mathbf{j} + \mathbf{k}) \text{ and } \mathbf{r} = 2\mathbf{i} - \mathbf{j} + 2\mathbf{k} + t(-\mathbf{i} - \mathbf{j} + \mathbf{k})$$

(c) Find the angle between the lines given in (b).

1 **r** = 5**i** − 2**j**
2 |**r**| = 13, 67.4° clockwise.
3 70.9°
4 (Answers are not unique, other answers are possible.)
(a) **r** = 3**i** + **k** + t(4**i** − **j** + 2**k**) (b) **r** = 4**i** − **j** + **k** + t(−7**i** + 3**j** − 3**k**)
5 (a) (1½, −1⅔) (b) Intersect at (7, 4, −3) (c) 121.5°

Core 3 and Core 4 (Pure Mathematics)

Sample questions and model answers

Core 4

1

(a) Express $f(x) = \dfrac{5x+1}{(x+2)(x-1)^2}$ in partial fractions.

(b) Hence show that $\displaystyle\int_{-1}^{0} f(x)\,dx = 1 - 2\ln 2$.

Remember the format when there is a repeated factor.

Take care with the denominator.

(a) Let $\dfrac{5x+1}{(x+2)(x-1)^2} \equiv \dfrac{A}{x+2} + \dfrac{B}{x-1} + \dfrac{C}{(x-1)^2}$

$\Rightarrow \dfrac{5x+1}{(x+2)(x-1)^2} \equiv \dfrac{A(x-1)^2 + B(x+2)(x-1) + C(x+2)}{(x+2)(x-1)^2}$

$\Rightarrow 5x + 1 \equiv A(x-1)^2 + B(x+2)(x-1) + C(x+2)$

Putting $x = 1$ makes the factor $(x - 1)$ equal to zero.

You could equate the x term or the constant term if you wish.

Let $x = 1$: $\quad 6 = 3C \Rightarrow C = 2$

Let $x = -2$: $\quad -9 = 9A \Rightarrow A = -1$

Equate x^2 terms: $\quad 0 = A + B \Rightarrow B = 1$

$\therefore f(x) = -\dfrac{1}{x+2} + \dfrac{1}{x-1} + \dfrac{2}{(x-1)^2}$

(b) $\displaystyle\int_{-1}^{0} f(x)\,dx = \int_{-1}^{0}\left(-\dfrac{1}{x+2} + \dfrac{1}{x-1} + \dfrac{2}{(x-1)^2}\right)dx$

Remember the modulus sign.

$= \Big[-\ln|x+2| + \ln|x-1| - 2(x-1)^{-1}\Big]_{-1}^{0}$

$= -\ln 2 + \ln|-1| + 2 - (-\ln 1 + \ln|-2| + 1)$

$= -\ln 2 + 2 - \ln 2 - 1$

$\ln 1 = \ln|-1| = 0$
$\ln|-2| = \ln 2.$

$= 1 - 2\ln 2$

Core 4

2

Expand $(1 - 2x)^{-3}$ in ascending powers of x as far as the term in x^3 and state the set of values of x for which the expansion is valid.

Take care with the negatives.

Remember to include the restriction on x, even when it is not requested specifically in the question.

$(1 - 2x)^{-3} = 1 + (-3)(-2x) + \dfrac{(-3)(-4)}{2!}(-2x)^2 + \dfrac{(-3)(-4)(-5)}{3!}(-2x)^3 + \ldots$

$= 1 + 6x + 24x^2 + 80x^3 + \ldots \quad$ provided $|2x| < 1$, i.e. $|x| < 0.5$

58

Core 3 and Core 4 (Pure Mathematics)

Sample questions and model answers (continued)

Core 4

3

Use the substitution $x = \tan u$ to show that $\int_0^1 \frac{1}{(1+x^2)^2}\,dx = \int_0^{\frac{\pi}{4}} \cos^2 u\,du$.

Hence show that $\int_0^1 \frac{1}{(1+x^2)^2}\,dx = \frac{\pi}{8} + \frac{1}{4}$.

Remember to change the x limits to u limits.

$$\int_0^1 \frac{1}{(1+x^2)^2}\,dx = \int_{x=0}^{x=1} \frac{1}{(1+x^2)^2}\frac{dx}{du}\,du$$

$$= \int_0^{\frac{\pi}{4}} \frac{1}{(\sec^2 u)^2}\sec^2 u\,du$$

$$= \int_0^{\frac{\pi}{4}} \frac{1}{\sec^2 u}\,du$$

$$= \int_0^{\frac{\pi}{4}} \cos^2 u\,du$$

Show this essential side working.

Let $x = \tan u$

$\Rightarrow \frac{dx}{du} = \sec^2 u$

$1 + x^2 = 1 + \tan^2 u = \sec^2 u$

Limits:
When $x = 0$, $\tan u = 0 \Rightarrow u = 0$
When $x = 1$, $\tan u = 1 \Rightarrow u = \frac{\pi}{4}$

To integrate even powers of sin x or cos x, change to double angles using cos 2x formula.

$$\int_0^{\frac{\pi}{4}} \cos^2 u\,du = \frac{1}{2}\int_0^{\frac{\pi}{4}} (1 + \cos 2u)\,du$$

$$= \frac{1}{2}\left[u + \frac{1}{2}\sin 2u\right]_0^{\frac{\pi}{4}}$$

$$= \frac{1}{2}\left(\frac{\pi}{4} + \frac{1}{2} - 0\right)$$

$$= \frac{\pi}{8} + \frac{1}{4}$$

$\cos 2u = 2\cos^2 u - 1$

$\Rightarrow \cos^2 u = \frac{1}{2}(1 + \cos 2u)$

Core 3

4

(a) Solve the inequality $|2x - 3| < 11$

(b) Sketch the graph of $y = |f(x)|$ where $f(x) = (x-1)(x-2)(x-3)$

(a) $|2x - 3| < 11$

This first step is important.

$\Rightarrow -11 < 2x - 3 < 11$

You can simplify the inequality by doing the same thing to each part.

$\Rightarrow -8 < 2x < 14$ — Add 3 to each part.

$\Rightarrow -4 < x < 7$. — Divide each part by 2.

(b)

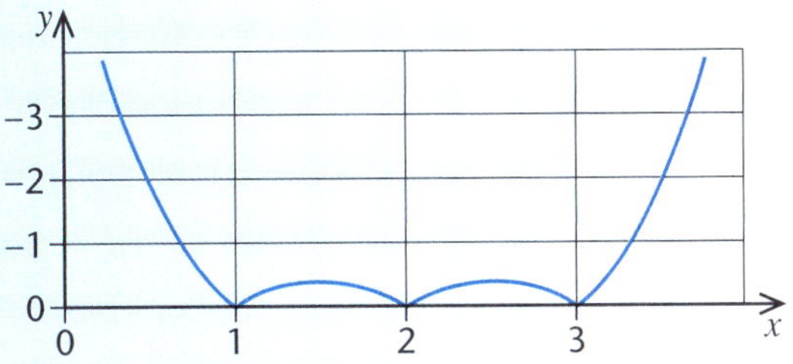

Parts of the graph of $y = f(x)$ that would be below the x-axis are reflected in the x-axis to give the graph of $y = |f(x)|$.

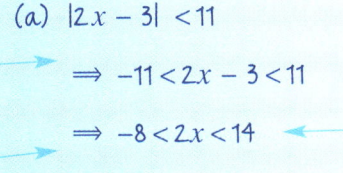

Core 3 and Core 4 (Pure Mathematics)

Sample questions and model answers (continued)

Core 3

5

The functions f and g are defined by:

$$f(x) = \sqrt{x-1} \quad x \geqslant 1$$
$$g(x) = 2x + 3 \quad x \in \mathbb{R}$$

(a) Find an expression for $f^{-1}(x)$ and state its domain and range.

(b) Find the value of:

  (i) $fg(7)$

  (ii) $gf(10)$

  (iii) $f^{-1}g^{-1}(15)$

Set $y = f(x)$ and rearrange to make x the subject.

(a) $y = \sqrt{x-1}$

$\Rightarrow y^2 = x - 1$

$\Rightarrow x = y^2 + 1$

$\Rightarrow f^{-1}(x) = x^2 + 1$

The domain of f^{-1} is given by the range of f.

The domain of f^{-1} is $x \geqslant 0$ and the range of f^{-1} is $x \geqslant 1$.

The range of f^{-1} is given by the domain of f.

The order in which you apply the functions is important.

(b) (i) $fg(7) = f(17) = 4$

  (ii) $gf(10) = g(3) = 9$

  (iii) $f^{-1}g^{-1}(15) = f^{-1}(6) = 37$

Core 4

6

The equation of a straight line l is

$\mathbf{r} = 4\mathbf{i} - 2\mathbf{k} + t(3\mathbf{i} + \mathbf{k})$ where t is a scalar parameter.

Points A, B and C lie on the line l. Referred to the origin O, they have position vectors $\mathbf{a}$, $\mathbf{b}$ and $\mathbf{c}$ respectively.

(a) Find $\mathbf{a}$ and $\mathbf{b}$ and angle AOB, given that distance $OA = OB = 10$ units.

(b) Find $\mathbf{c}$, given that OC is perpendicular to l.

(a) $\mathbf{a} = (4 + 3t)\mathbf{i} + (-2 + t)\mathbf{k}$ for some value t.   (1)

Distance $OA = |\mathbf{a}|$.

Since $|\mathbf{a}| = 10$,

$\sqrt{(4+3t)^2 + (-2+t)^2} = 10$

$16 + 24t + 9t^2 + 4 - 4t + t^2 = 100$

$10t^2 + 20t - 80 = 0$

$t^2 + 2t - 8 = 0$

$(t-2)(t+4) = 0$

$\Rightarrow t = 2$ or $t = -4$

Core 3 and Core 4 (Pure Mathematics)

Sample questions and model answers (continued)

*This gives the position vectors of **a** and **b**.*

Substituting for t in (1) gives
$$\underline{a} = 10\underline{i} \quad \text{and} \quad \underline{b} = -8\underline{i} - 6\underline{k}$$

Let angle AOB be θ.

$$|\underline{a}| = |\underline{b}| = 10 \quad \text{and} \quad \underline{a} \cdot \underline{b} = (10\underline{i}) \cdot (-8\underline{i} - 6\underline{k}) = -80$$

$$\underline{a} \cdot \underline{b} = |\underline{a}||\underline{b}|\cos\theta$$

$$\Rightarrow \cos\theta = \frac{-80}{100} = -0.8$$

$$\Rightarrow \theta = 143.1° \text{ (1 d.p.)}$$

(b) $\underline{c} = (4 + 3t)\underline{i} + (-2 + t)\underline{k}$ for some value t. (2)

$3\mathbf{i} + \mathbf{k}$ is the direction vector of l and the scalar product of perpendicular vectors is zero.

Since $\underline{c}$ is perpendicular to l,

$$((4 + 3t)\underline{i} + (-2 + t)\underline{k}) \cdot (3\underline{i} + \underline{k}) = 0$$

$$\Rightarrow 12 + 9t - 2 + t = 0$$

$$\Rightarrow t = -1$$

Substituting for t in (2) gives $\underline{c} = \underline{i} - 3\underline{k}$.

Core 3 and Core 4 (Pure Mathematics)

Practice examination questions

Core 4

1. It is given that $f(x) = \dfrac{1}{(1+x)^2} + \sqrt{4+x}$.

   Show that if x^3 and higher powers are ignored, $f(x) \approx a + bx + cx^2$ and find the values of a, b and c.

   State the values of x for which the expansion is valid.

Core 4

2. (a) Find $\dfrac{dy}{dx}$ when $y = \cos^2 x \sin x$, expressing your answer in terms of $\cos x$.

   (b) Using the substitution $u = \cos x$, or otherwise, find $\displaystyle\int \cos^2 x \sin x \, dx$.

Core 3

3. A curve has equation $y = \dfrac{x^2 - 4}{x + 1}$.

   The equation of the normal to the curve at $(2, 0)$ is $ax + by + c = 0$, where a, b and c are integers.

   Find the values of a, b, and c.

Core 3

4. The functions f and g are defined by:

   $f: x \mapsto 5x - 7 \quad x \in \mathbb{R}$

   $g: x \mapsto (x+1)(x-1) \quad x \in \mathbb{R}$.

   (a) Find the range of g.

   (b) Find an expression for $fg(x)$.

   (c) Determine the values of x for which $fg(x) = f(x)$.

Core 3

5. (a) Express $15 \sin x + 8 \cos x$ in the form $R \sin(x + \alpha)$ where $0 < \alpha < 90°$.

   (b) State the maximum value of $15 \sin x + 8 \cos x$.

   (c) Solve the equation $15 \sin x + 8 \cos x = 12$.

   Give all the solutions in the interval $0 < x < 360°$.

Core 4

6. Expressing y in terms of x, solve the differential equation

   $$x \dfrac{dy}{dx} = y + yx,$$

   given that $y = 4$ when $x = 2$.

Core 3

7. (a) Use the chain rule to differentiate:

   (i) $y = (4x - 3)^9$

   (ii) $y = \ln(5 - 2x)$

   (b) Using your results from part (a):

   (i) Write down an expression for $\displaystyle\int (4x - 3)^8 \, dx$

   (ii) Calculate $\displaystyle\int_3^4 \dfrac{1}{5 - 2x} \, dx$

62

Core 3 and Core 4 (Pure Mathematics)

Practice examination questions (continued)

Core 4

8 (a) Use integration by parts to evaluate $\int_0^1 x e^{-2x} dx$.

(b) The diagram shows the graph of $y = x e^{-x}$.

The curve has a maximum point at A.

The shaded region R is the area bounded by the curve, the x-axis and the vertical line through A.

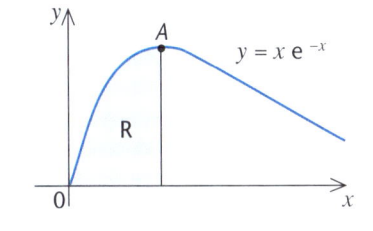

(i) Show that the coordinates of A are $(1, e^{-1})$.
(ii) The region R is rotated completely about the x-axis.
Calculate the volume of the solid generated.

Core 4

9 In a biology experiment, the growth of a population is being investigated.

A simple model is set up in which the rate of increase of the population at time t is proportional to the number, P, in the population at that time.

It is known that when $t = 0$, $P = 500$ and when $t = 10$, $P = 1000$.

Obtain the relationship between P and t and estimate the size of the population, according to the model, when $t = 20$.

Core 4

10 (a) Find the equation of the normal to the curve $x^3 + xy + y^2 = 7$ at the point $(1, 2)$.

(b) Find the coordinates of the two stationary points on the curve $x = t^2$, $y = t + \frac{1}{t}$.

Core 4

11 The region R, shaded in the diagram, is bounded by the curve $y = \sin 2x$ and the x-axis.

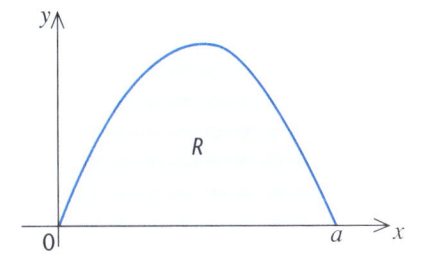

(a) Find the value of a.
(b) Show that the area of R is 1 square unit.
(c) Find the volume of the solid formed when R is rotated completely about the x-axis.

Core 4

12 (a) Find a vector equation of the line AB through $A(2, 2, 0)$ and $B(2, 3, 1)$.

(b) Show that the line AB and the line $\mathbf{r} = 2\mathbf{j} + \mathbf{k} + \lambda(\mathbf{i} + \mathbf{k})$, where λ is a scalar parameter, have no common point.

(c) Find the acute angle between the two lines.

Chapter 2
Probability and Statistics 2

The following topics are covered in this chapter:

- Continuous random variables
- The Poisson distribution and approximations
- Estimation and sampling
- Hypothesis tests

2.1 Continuous random variables

LEARNING SUMMARY

After studying this section you should be able to:

- use the normal distribution model to calculate probabilities
- find probabilities and the mean and variance of a continuous variable with probability density function $f(x)$

The normal distribution

The **normal variable** is a continuous random variable. Its probability function is represented by a bell-shaped curve, symmetric about the mean value μ.

The parameters of the normal distribution are μ and σ^2. If the continuous random variable X follows a normal distribution with these **parameters** then this is written as $X \sim N(\mu, \sigma^2)$. The standard deviation is σ.

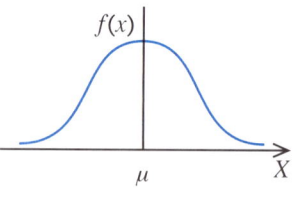

> Tabulated values of probabilities for the normal distribution use Z as the variable.

The **standard normal variable** Z is used to calculate probabilities based on the Normal distribution where $Z = \dfrac{X - \mu}{\sigma}$ and $Z \sim N(0, 1)$.

Example
Find $P(15 < X < 17)$ where $X \sim N(15, 25)$

$$15 < X < 17 \Rightarrow \frac{15-15}{5} < Z < \frac{17-15}{5} \Rightarrow 0 < Z < 0.4$$

$$P(15 < X < 17) = P(0 < Z < 0.4)$$
$$= \Phi(0.4) - \Phi(0) = 0.6554 - 0.5000$$
$$= 0.1554$$

> Using the normal distribution tables.

You may need to use given information to set up and solve simultaneous equations to find the values of μ and σ.

Example
It is given that $P(X > 70) = 0.0436$ and $P(X < 45) = 0.3669$, where $X \sim N(\mu, \sigma^2)$. Find the values of μ and σ.

$$P(X > 70) = P\left(Z > \frac{70-\mu}{\sigma}\right) = 0.0436$$

$$\Phi\left(\frac{70-\mu}{\sigma}\right) = 1 - 0.0436 = 0.9564$$

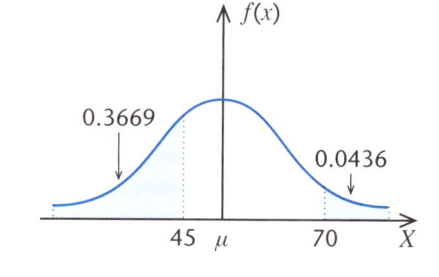

Probability and Statistics 2

From the tables

$$\left(\frac{70-\mu}{\sigma}\right) = 1.71 \Rightarrow 70 - \mu = 1.71\sigma \quad [1]$$

Similarly

$$P(X < 45) = P\left(Z < \frac{45-\mu}{\sigma}\right) = 0.3669$$

$$\Phi\left(\frac{45-\mu}{\sigma}\right) = 0.3669$$

In general, $\Phi(-z) = 1 - \Phi(z)$.

$$\Phi\left(-\left(\frac{45-\mu}{\sigma}\right)\right) = 1 - 0.3669 = 0.6331$$

$$-\left(\frac{45-\mu}{\sigma}\right) = 0.34$$

$$45 - \mu = -0.34\sigma \quad [2]$$

[1] − [2] gives $25 = 2.05\sigma \Rightarrow \sigma = 12.2$

Substituting in (1): $\mu = 49.1$

The probability density function, $f(x)$

The continuous random variable X is defined by its **probability density function** $f(x)$, together with the values for which it is valid.

Probabilities are given by areas under the curve.

$$P(c \leq X \leq d) = \int_c^d f(x)\,dx$$

The total probability is 1.

If $f(x)$ is valid for $a \leq x \leq b$, then $\int_a^b f(x)\,dx = 1$.

The probability of obtaining a specific value is zero.

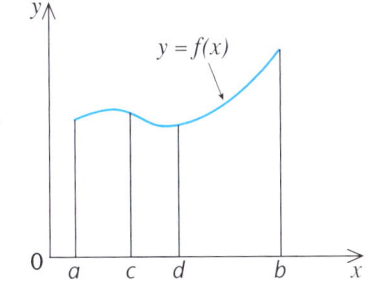

$$P(X = c) = 0$$

The mean is the **expectation** of X, written $E(X)$. It is often denoted by μ.

If $f(x)$ is given in stages, then the integration is carried out in stages also.

KEY POINT

Mean: $\quad E(X) = \mu = \int_{\text{all } x} x f(x)\,dx$

Variance: $\quad \text{Var}(X) = \sigma^2 = E(X^2) - \mu^2$, where $E(X^2) = \int_{\text{all } x} x^2 f(x)\,dx$

Example

The continuous random variable X has probability density function $f(x)$ given by $f(x) = \frac{1}{4}x$ for $1 \leq x \leq 3$. Find the mean μ, $E(X^2)$ and the standard deviation σ.

$$\mu = E(X) = \int_{\text{all } x} x f(x)\,dx = \int_1^3 \frac{1}{4} x^2 \,dx = \left[\frac{x^3}{12}\right]_1^3 = 2\frac{1}{6}$$

$$E(X^2) = \int_{\text{all } x} x^2 f(x)\,dx = \int_1^3 \frac{1}{4} x^3 \,dx = \left[\frac{x^4}{16}\right]_1^3 = 5$$

$\sigma = \sqrt{\text{variance}}$

$$\text{Var}(X) = 5 - (2\tfrac{1}{6})^2 = \tfrac{11}{36} \Rightarrow \sigma = \sqrt{\tfrac{11}{36}} = 0.553 \text{ (3 d.p.)}$$

Probability and Statistics 2

Progress check

1. Find $P(17 < X < 24)$ where $X \sim N(14, 16)$.

2. It is given that $X \sim N(\mu, \sigma^2)$ and $P(X < 30) = 0.7$
   Given also that $P(X > 35) = 0.1$, find μ and σ.

3. $f(x) = 1 - 0.5x$, $0 \leqslant x \leqslant 2$.
   Find $E(X)$, $E(X^2)$ and $Var(X)$.

4. The continuous random variable X has probability density function $f(x) = 0.25$ for $3 \leqslant x \leqslant a$ and $f(x) = 0$ otherwise.
   (a) Find the value of a.
   (b) Write down the value of the median of X.
   (c) Calculate the standard deviation of X.

---

1. 0.2204
2. $\mu = 26.5$
   $\sigma = 6.60$
3. $\frac{2}{3}, \frac{2}{3}, \frac{2}{9}$
4. (a) 7 (b) 5 (c) 1.15 (3 s.f.)

Probability and Statistics 2

2.2 The Poisson distribution and approximations

After studying this section you should be able to:

- understand the Poisson distribution
- use the Poisson approximation to the binomial distribution
- use the normal approximation to the binomial distribution
- use the normal approximation to the Poisson distribution

LEARNING SUMMARY

The Poisson distribution

The **Poisson distribution** is used to model the number of occurrences, X, of an event in a given interval of space or time, when the events occur at a constant average rate, independently and at random.

λ is the **parameter** of the distribution.

If the mean number of occurrences in the interval is λ, then $X \sim \text{Po}(\lambda)$.

KEY POINT

If $X \sim \text{Po}(\lambda)$:
$$P(X = x) = e^{-\lambda} \frac{\lambda^x}{x!} \quad \text{for } x = 0, 1, 2, 3, \ldots$$

Mean, $\mu = \lambda$

Mean = variance

Variance $= \lambda$ and standard deviation $= \sqrt{\lambda}$

This is similar to using cumulative Binomial tables in Probability and Statistics 1.

For some values of λ you can use **cumulative Poisson probability tables** to find $P(X \leq x)$ and from this you can find other probabilities, for example:

$P(X < 2) = P(X \leq 1)$ **0 1** 2 3 …

$P(X > 4) = 1 - P(X \leq 4)$ 0 1 2 3 4 **5 6 7** …

$P(X \geq 3) = 1 - P(X \leq 2)$ 0 1 2 **3 4 5** …

$P(X = 4) = P(X \leq 4) - P(X \leq 3)$ 0 1 2 3 **4** 5 …

Example

The number of letters received by the Adams family on a weekday may be modelled by a Poisson distribution with mean 3.75 and the number of letters received on a Saturday may be modelled by a Poisson distribution with mean 3. Determine the probability that:

(a) 4 letters are received on Monday

(b) more than 5 letters are received on Saturday.

$\lambda = 3.75$ is not in the Poisson tables so you have to use the formula.

(a) X is the number of letters received on Monday.

$X \sim \text{Po}(3.75)$ so $P(X = 4) = e^{-3.75} \dfrac{3.75^4}{4!} = 0.1937\ldots = 0.194$ (3 s.f.)

$\lambda = 3$ is in the Poisson tables.

(b) Y is the number of letters received on Saturday.

$Y \sim \text{Po}(3)$ so $P(Y > 5) = 1 - P(Y \leq 5) = 1 - 0.9161 = 0.0839$

67

Probability and Statistics 2

The Poisson approximation to the binomial distribution

The binomial distribution $X \sim B(n,p)$ is used to model the number of successes in n independent trials when the probability of success, p, is constant.

The mean of the binomial distribution is np and the variance is npq (where $q = 1 - p$).

> With these conditions, $np \approx npq$. This fits in with the Poisson distribution where the mean and the variance are equal.

KEY POINT

When n is large ($n > 50$) and $np < 5$, $X \sim B(n,p)$ can be **approximated** by a Poisson distribution with the same mean, where $X \sim Po(np)$ approximately.

Example

The proportion of defective items produced by a machine is 2.5%. Find the probability of obtaining fewer than 3 defective items in a random sample of 100 items.

If X is the number of defective items in 100, then $X \sim B(100, 0.025)$.
The mean $np = 100 \times 0.025 = 2.5$
Since $n > 50$ and $np < 5$, use the Poisson approximation, $X \sim Po(2.5)$.

> $P(X = x) = \dfrac{\lambda^x}{x!} e^{-\lambda}$

$P(X < 3) = P(X = 0) + P(X = 1) + P(X = 2)$

$= e^{-2.5} + 2.5e^{-2.5} + \dfrac{2.5^2}{2!} e^{-2.5} = e^{-2.5}\left(1 + 2.5 + \dfrac{2.5^2}{2!}\right) = 0.54$ (2 s.f.)

> It is often quicker to use cumulative probability tables, so make sure that you know how to use them.

If you have access to cumulative Poisson tables giving $P(X \leq r)$ for various values of r, then these tables can be used with $\lambda = 2.5$ and
$P(X < 3) = P(X \leq 2) = 0.5438 = 0.54$ (2 s.f.).

The normal approximation to the binomial distribution

KEY POINT

When n is large enough to ensure that $np > 5$ and $nq > 5$ (where $q = 1 - p$), $X \sim B(n,p)$ can be **approximated** by a normal distribution with the same mean and the same variance, where $X \sim N(np, npq)$ approximately.

The larger the value of n and the closer that p is to 0.5, the better the approximation.

The normal distribution is continuous, whereas the binomial is discrete, so a **continuity correction** must be used.

Example

The probability that a person supports a particular organisation is 0.3. In a random sample of 50 people, find the probability that at least 20 support the organisation.

> Always define the variable, giving its distribution if known.

If X is the number of people in 50 who support the organisation, then $X \sim B(50, 0.3)$.

Mean $np = 50 \times 0.3 = 15$, variance $npq = 50 \times 0.3 \times 0.7 = 10.5$
Conditions for normal approximation: $np = 15 > 5$, $nq = 50 \times 0.7 = 35 > 5$

> $X \sim N(\mu, \sigma^2)$ with $\mu = 15$ and $\sigma^2 = 10.5 \Rightarrow \sigma = \sqrt{10.5}$

$\therefore \quad X \sim N(15, 10.5)$ approximately.

Probability and Statistics 2

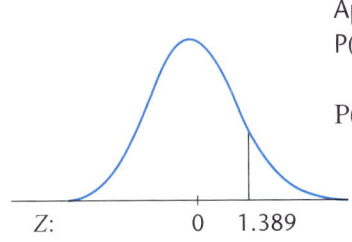

Applying the continuity correction,
$P(X \geq 20)$ becomes $P(X > 19.5)$.

$$P(X > 19.5) = P\left(Z > \frac{19.5 - 15}{\sqrt{10.5}}\right)$$
$$= P(Z > 1.389)$$
$$= 1 - \Phi(1.389) = 1 - 0.9176 = 0.082 \text{ (2 s.f.)}$$

Z: 0   1.389

Think of $X = 20$ as going from 19.5 to 20.5. Since 'at least 20' includes $X = 20$, find $P(X > 19.5)$ in the normal distribution.

Standardise
$$Z = \frac{X - \mu}{\sigma}$$
and then use standard normal tables.

The normal approximation to the Poisson distribution

The Poisson distribution, $X \sim \text{Po}(\lambda)$, is used to model the number of occurrences of an event, when events occur randomly. The mean is λ and the variance is λ.

> **KEY POINT**
> When λ is large ($\lambda > 15$ say), $X \sim \text{Po}(\lambda)$ can be approximated by a normal distribution with the same mean and variance, where $X \sim N(\lambda, \lambda)$ approximately.

Since the Poisson distribution is discrete, a continuity correction must be used.

Example

If $X \sim \text{Po}(20)$, find $P(15 < X < 22)$.

Since n is large, $X \sim N(20, 20)$ approximately.

$P(15 < X < 22)$ becomes $P(15.5 < X < 21.5)$.

$$P(15.5 < X < 21.5) = P\left(\frac{15.5 - 20}{\sqrt{20}} < Z < \frac{21.5 - 20}{\sqrt{20}}\right)$$
$$= P(-1.006 < Z < 0.335)$$
$$= \Phi(1.006) + \Phi(0.335) - 1$$
$$= 0.8427 + 0.6312 - 1$$
$$= 0.474 \text{ (3 s.f.)}$$

You do not want to include 15 or 22 so go from 15.5 to 21.5.

Standardise the variables and then use normal tables.

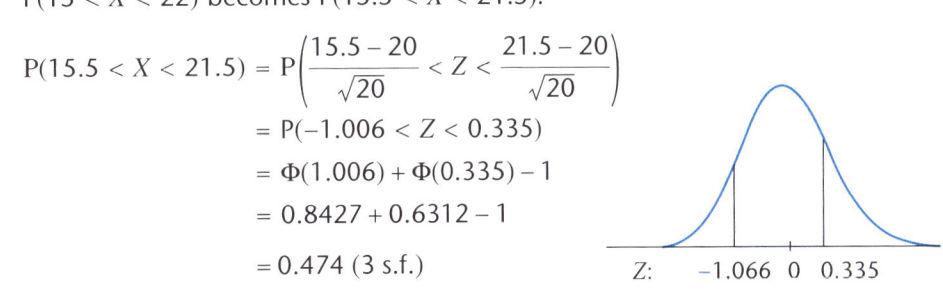

Apply the continuity correction.

Z: −1.066 0 0.335

Probability and Statistics 2

Progress check

1. The variable X follows a Poisson distribution with standard deviation 2.
   (a) Write down the value of λ.
   (b) Use cumulative probability tables to find
   (i) $P(X < 5)$ (ii) $P(X \geq 1)$ (iii) $P(X = 6)$

2. The number of telephone calls to a switchboard may be modelled by a Poisson distribution with mean 2.75 calls per 5-minute interval. Determine, to 3 significant figures, the probability that in a five-minute interval,
   (a) there will be at most one call
   (b) there will be exactly 5 calls.

3. $X \sim B(80, 0.06)$. Find, to 3 decimal places, $P(X = 4)$
   (a) by using the binomial distribution
   (b) by using a suitable approximation.

4. A newspaper reports that 62% of adults have an e-mail address. A random sample of 40 adults was selected. Using a suitable approximation, find the probability that at least 30 had an e-mail address.

5. Telephone calls reach a switchboard independently and at random at a rate of 30 per hour. Find the probability that in a randomly selected period of one hour there are fewer than 20 calls.

1 (a) 4 (b)(i) 0.6288 (ii) 0.9817 (iii) 0.1042
2 (a) 0.240 (3 s.f.) (b) 0.0838 (3 s.f.)
3 (a) 0.186 (b) 0.182
4 0.063 (2 s.f.)
5 0.028 (2 s.f.)

Probability and Statistics 2

2.3 Estimation and sampling

After studying this section you should be able to:

- distinguish between different sampling techniques
- find unbiased estimates for population parameters from a sample
- understand the theory relating to sampling distributions
- use the central limit theorem

LEARNING SUMMARY

Sampling techniques

When you want information about a **population**, you could carry out a **census** of every member. The advantage is that you would have accurate and complete information. Disadvantages include cost and time requirements. Taking a census could also destroy the population. For example, if you were investigating the length of life of a calculator battery, carrying out a census would destroy the population.

More usually, a **sample** is taken. The sample should be representative of the whole population, so **bias** in the choice of sample members must be avoided.

In a **random sample** of size n, each member of the population has an equal chance of being selected. Also all subsets of the population of size n have an equal chance of constituting the sample.

Assigning a number to each member of a population and then drawing the numbers out of a hat is one method of obtaining a **simple random sample**. Another is to use **random number tables**, in which each digit has an equal chance of occurring.

> If there is a periodic fault, then systematic sampling may fail to register it or may give it undue weight.

Random sampling from a very large population can be very laborious. It may be more convenient to carry out **systematic sampling**. This involves selecting every kth member of the population, for example, every 10th item from a particular machine on a production line.

> The stratified sample can be constructed in proportion to the number of members in each stratum.

The method of **stratified sampling** is often used when the population is split into distinguishable layers or strata, such as students in each faculty in a college.

Unbiased estimates

When a **population parameter**, such as the mean or the variance, is **unknown**, then it is sensible to estimate it from a sample.

An **unbiased estimate** is one which, on the average, gives the true value, i.e. E(estimate) = true value of parameter.

> The best unbiased estimate is the estimate with the smallest variance.

Best unbiased estimates:

For mean μ $\qquad \hat{\mu} = \bar{x} = \dfrac{\sum x}{n} \qquad \bar{x}$ is sample mean

For variance $\sigma^2 \qquad \hat{\sigma}^2 = \dfrac{n}{n-1}\left(\dfrac{\sum x^2}{n} - \left(\dfrac{\sum x}{n}\right)^2\right)$

Sometimes S^2 is used for $\hat{\sigma}^2$

Alternative formats: $S^2 = \hat{\sigma}^2 = \dfrac{\sum(x-\bar{x})^2}{n-1}$, $S^2 = \hat{\sigma}^2 = \dfrac{1}{n-1}\left(\sum x^2 - \dfrac{(\sum x)^2}{n}\right)$.

> **Key points from AS**
> - Mean, variance and standard deviation
>   *Revise AS pages 75–77*
>
> Make sure that you are familiar with the format given in your examination booklet and practise using it.

Probability and Statistics 2

> If you are given only summary data such as Σx or Σx^2, substitute into the appropriate formula.

A calculator in statistical mode can be used to obtain the value of $\hat{\sigma}$ directly from raw data. Look for the key marked $x\sigma n - 1$.

Sampling distributions

The most important sampling distribution is the distribution of the sample mean and you will need to be able to use it to find probabilities.

The sampling distribution of the mean

If all possible samples of size n are taken from $X \sim N(\mu, \sigma^2)$ and their sample means calculated, then these means form the sampling distribution of means, $\bar{X}$ which is also normally distributed.

KEY POINT
If $X \sim N(\mu, \sigma^2)$, then $\bar{X} \sim N\left(\mu, \dfrac{\sigma^2}{n}\right)$.

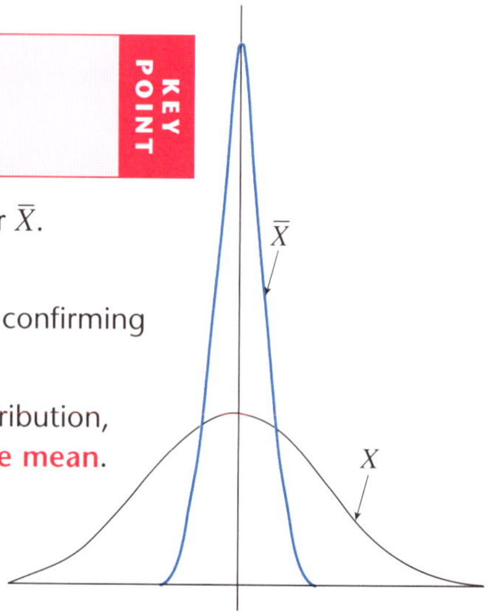

> The mean of $\bar{X}$ is the same as the mean of X.
>
> The variance of $\bar{X}$ is much smaller than the variance of X.
>
> Standard deviation = $\sqrt{\text{variance}}$

The diagram shows the curves for X and for $\bar{X}$.

They are both symmetrical about μ.
The curve for $\bar{X}$ is much more squashed in, confirming the smaller standard deviation.

The standard deviation of the sampling distribution, $\sigma/\sqrt{n}$, is known as the **standard error of the mean**.

The following result is extremely useful and should be learnt.

KEY POINT
If the distribution of X is not normal, then, by the **central limit theorem**, $\bar{X} \sim N\left(\mu, \dfrac{\sigma^2}{n}\right)$ approximately, **provided n is large**.

> $\hat{\sigma}^2$ is the unbiased estimate of σ^2.

If you do not know σ^2, the variance of X, then it can be estimated by $\hat{\sigma}^2$ and provided n is large, $\bar{X} \sim N\left(\mu, \dfrac{\hat{\sigma}^2}{n}\right)$ approximately.

Example

The heights of men in a particular area are normally distributed with mean 176 cm and standard deviation 7 cm. A random sample of 50 men is taken. Find the probability that the mean height of the men in the sample is less than 175 cm.

If X is the height, in centimetres, of a man from this area, then $X \sim N(176, 7^2)$.

For random samples of size 50, $\bar{X} \sim N\left(176, \dfrac{7^2}{50}\right)$, i.e. $\bar{X} \sim N(176, 0.98)$.

> In this example, since the population is normal, any size sample, large or small, could have been taken.

$$P(\bar{X} < 175) = P\left(Z < \dfrac{175 - 176}{\sqrt{0.98}}\right)$$
$$= P(Z < -1.010\ldots)$$
$$= 1 - \Phi(1.010)$$
$$= 1 - 0.8438$$
$$= 0.156 \text{ (3 d.p.)}$$

$\bar{X}$: 175 176
Z: -1.010 0

72

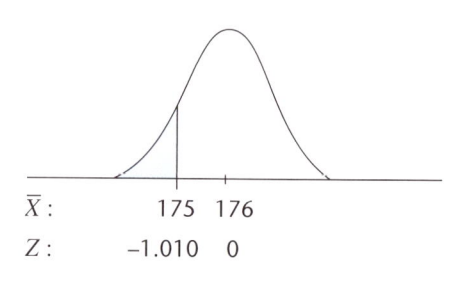

Probability and Statistics 2

Progress check

1. The random variable X has mean μ and standard deviation σ.
   Ten independent observations of X are
   4.6, 3.9, 4.2, 5.8, 6.3, 4.2, 7.2, 6.0, 7.1, 5.4.

   Find unbiased estimates of μ and σ.

2. A random sample of size 100 is taken from a normal distribution with mean 120 and standard deviation 5.
   (a) Find P(118.75 < $\bar{X}$ < 120.5), where $\bar{X}$ is the sample mean.
   (b) State, with a reason, whether your answer would be different if X is not normally distributed.

1 5.47, 1.21
2 (a) 0.8351   (b) no, use central limit theorem since n is large

Probability and Statistics 2

2.4 Hypothesis tests

LEARNING SUMMARY

After studying this section you should be able to:

- understand the language used in hypothesis testing
- understand Type I and Type II errors
- understand how to obtain critical z-values
- perform the z-test for the mean
- perform tests for a binomial proportion (large and small sample sizes)
- perform a discrete test for a Poisson mean

The language of hypothesis testing

When you want to know something about a population, you might perform a **hypothesis** or **significance test**.

A **null hypothesis**, H_0 is made about the population.
Then the **alternative hypothesis** H_1, the situation when H_0 is not true, is stated.

Example

When investigating the mean of a normal distribution, you might make the null hypothesis that the mean is 5. This is written $H_0: \mu = 5$.

The alternative hypothesis would be one of the following:

- $H_1: \mu > 5$ (one-tailed, upper tail test)
- $H_1: \mu < 5$ (one-tailed, lower tail test)
- $H_1: \mu \neq 5$ (two-tailed test).

> Use $\mu > 5$ if you suspect an *increase*.
>
> Use $\mu \neq 5$ if you suspect a *change*.

> Use $\mu < 5$ if you suspect a *decrease*.

> The formula for the test statistic depends on the sampling distribution. Examples are shown in the following text.

A **test statistic** is defined and its distribution when H_0 is true is stated.
Its value, based on information from a random sample taken from the population, is found. This is the **test value**.

The hypothesis test involves deciding whether or not the test value could have come from the distribution defined by the null hypothesis.
If the test value is in the main bulk of the distribution (the **acceptance region**) it is *likely* to have come from the distribution. If it is in the 'tail' end (the **rejection** or **critical region**), it is *unlikely* to have come from the distribution.

> Probability theory is used to decide the placing of the boundary between 'likely' and 'unlikely'.

The decision rule (**rejection criterion**) is based on the **significance level** of the test, which is used to fix the boundaries for the rejection region. The boundaries are called **critical values**.

The significance level is a given as a percentage.

> The levels used most often are 10%, 5% and 1%.

For **example**, for a significance level of 5%, the critical values are such that 5% of the distribution is in the critical (rejection) region,

i.e. $P(X > \text{critical value}) = 0.05$, for a one-tailed, upper tail test
$P(X < \text{critical value}) = 0.05$, for a one-tailed, lower tail test
$P(X > \text{upper critical value}) = 0.025$, for a two-tailed test
$P(X < \text{lower critical value}) = 0.025$, for a two-tailed test.

> There are two critical values in a two-tailed test.

> A test statistic is said to be significant if it lies in the critical region. If it lies in the acceptance region, then it is not significant.

Depending on the position of the test value, the **decision** is made.
If the test value lies in the critical region, H_0 is rejected in favour of H_1.
If the test value is in the acceptance region, H_0 is not rejected.

The **conclusion** is then stated in relation to the situation being tested.

Probability and Statistics 2

Type I and Type II errors

> You make a Type I error when you reject a true null hypothesis.

In a hypothesis test, either H_0 is rejected or H_0 is not rejected. There are occasions when the decision is incorrect and an error is made.

If H_0 is rejected when it is in fact true, a **Type I** error is made.

The probability of making a Type I error is the same as the significance level of the test. For example, if the significance level is 5%, then P(Type I error) = 0.05

> You make a Type II error when you accept a false null hypothesis.

If H_0 is accepted (i.e. not rejected) when it is in fact false, a **Type II** error is made.

To find the probability of making a Type II error, a specific value for H_1 must be given. Then P(Type II error) = P(H_0 is accepted when H_1 is true).

Critical values for z-tests

Some hypothesis tests involving the normal distribution are known as z-tests. In a z-test, the critical values for the rejection region can be found using the standard normal distribution, Z. The most commonly used values are summarised below.

> 5%
> For one-tailed upper tail, $\Phi(z) = 0.95$
> For two tailed, upper tail, $\Phi(z) = 0.975$
> Use symmetry for lower tails.
>
> 10%
> For one-tailed upper tail, $\Phi(z) = 0.90$
> For two tailed, upper tail $\Phi(z) = 0.95$
>
> 1%
> For one-tailed upper tail, $\Phi(z) = 0.99$
> For two tailed, upper tail $\Phi(z) = 0.995$

Level	Type of test	Critical value	Decision rule
5%	One-tailed (upper)	1.645	Reject H_0 if $z > 1.645$
	One-tailed (lower)	−1.645	Reject H_0 if $z < -1.645$
	Two-tailed	±1.96	Reject H_0 if $z > 1.96$ or $z < -1.96$
10%	One-tailed (upper)	1.282	Reject H_0 if $z > 1.282$
	One-tailed (lower)	−1.282	Reject H_0 if $z < -1.282$
	Two-tailed	±1.645	Reject H_0 if $z > 1.645$ or $z < -1.645$
1%	One-tailed (upper)	2.326	Reject H_0 if $z > 2.326$
	One-tailed (lower)	−2.326	Reject H_0 if $z < -2.326$
	Two-tailed	±2.576	Reject H_0 if $z > 2.576$ or $z < -2.576$

z-test for the mean

This is used to test the mean μ of a population when you know the variance σ^2. If the population is **normal**, the samples can be **any size**, but if the population is **not normal**, then **large** samples must be taken, so that the central limit theorem can be applied.

The distribution considered for the test statistic is the sampling distribution of means (see page 72).

> When the population is not normal, the central limit theorem is needed.

KEY POINT

$$\text{Test statistic } Z = \frac{\bar{X} - \mu}{\sigma/\sqrt{n}} \sim N(0, 1).$$

This formula can also be used if the variance of X is not known, provided n is large. In this case, the unbiased estimate $\hat{\sigma}^2$ is used for σ^2.

Probability and Statistics 2

Example

A machine fills tins of baked beans such that the mass of a filled tin is normally distributed with mean mass 429 g and the standard deviation is 3 g. The machine breaks down and after being repaired, the mean mass of a random sample of 100 tins is found to be 428.45 g. Is this evidence, at the 5% level, that the mean mass of tins filled by this machine has decreased? Assume that the standard deviation remains unaltered.

Define your variables.

Let X be the mass, in grams, of a filled tin and let the mean after the repair be μ.

State the hypotheses.

$H_0: \mu = 429$; $H_1: \mu < 429$.

State the distribution of the test statistic assuming the null hypothesis is true.

Consider the sampling distribution of means $\bar{X} \sim N\left(429, \dfrac{3^2}{100}\right)$.

State the type of test and the rejection criterion.

Perform a one-tailed (lower tail) test at the 5% level and reject H_0 if $z < -1.645$

Calculate the value of the test statistic.

$\bar{x} = 428.45$, so
$$z = \dfrac{\bar{x} - \mu}{\sigma/\sqrt{n}} = \dfrac{428.45 - 429}{3/\sqrt{100}} = -1.833\ldots$$

Decide whether to reject H_0 or not and relate your conclusion to the question.

Since $z < -1.645$, the test value lies in the critical region and H_0 is rejected.

There is evidence, at the 5% level, that the **mean** mass of a filled tin has decreased.

Alternative method

Alternatively, a **probability method** can be used.
The rejection rule is to reject H_0 if $P(\bar{X} < 428.45) < 0.05$

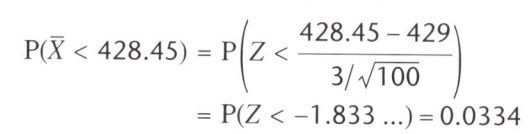

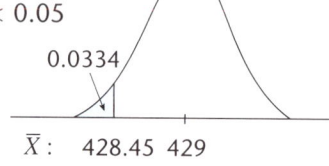

$$P(\bar{X} < 428.45) = P\left(Z < \dfrac{428.45 - 429}{3/\sqrt{100}}\right)$$
$$= P(Z < -1.833\ldots) = 0.0334$$

Since $P(\bar{X} < 428.45) < 0.05$, H_0 is rejected and the conclusion is as above.

Tests for a binomial proportion

This binomial test is used to test the proportion of successes, p, in a binomial population $X \sim B(n, p)$.

Large sample size

Normal approximation to the binomial, see page 68.

When n is large (such that $np > 5$ and $nq > 5$) a normal approximation to the binomial distribution can be used.

When testing in the upper tail, subtract 0.5, and when testing in the lower tail, add 0.5.

A continuity correction of $+0.5$ or -0.5 is needed as the binomial distribution is discrete and the normal distribution is continuous.

> **KEY POINT**
> Test statistic $Z = \dfrac{(X \pm 0.5) - np}{\sqrt{npq}} \sim N(0, 1)$, provided n is large.

The hypothesis test follows the same pattern as the z-test for the mean.

Example

A coin is tossed 50 times and 30 heads are obtained. Is this evidence, at the 5% level of significance, that the coin is biased in favour of heads?

Define the variable.

Let the probability that the coin shows heads be p.
If X is the number of heads in 50 tosses, then $X \sim B(50, p)$.

Probability and Statistics 2

State the hypotheses and also the distribution if the null hypothesis is true.

$H_0: p = 0.5$ (the coin is fair), $H_1: p > 0.5$ (the coin is biased in favour of heads).
If the null hypothesis is true, then $X \sim B(50, 0.5)$.

Justify the normal approximation.

But n is large such that $np = 50 \times 0.5 = 25 > 5$ and $nq = 25 > 5$, so $X \sim N(np, npq)$ approximately. Since $npq = 50 \times 0.5 \times 0.5 = 12.5$, $X \sim N(25, 12.5)$.

State the rejection criterion.

At the 5% level, reject H_0 if $z > 1.645$

29.5 is used because you need to test whether the whole rectangle representing $x = 30$ (from 29.5 to 30.5) is in the upper tail critical region.

The test value is $x = 30$, so using the continuity correction,
$$z = \frac{(x - 0.5) - np}{\sqrt{npq}} = \frac{29.5 - 25}{\sqrt{12.5}} = 1.27 \ldots$$

Since $z < 1.645$, the test value does not lie in the critical region and H_0 is not rejected.

Relate the conclusion to the question.

At the 5% level there is not sufficient evidence to say that the coin is biased in favour of heads.

Binomial test when the sample size is small

When n is small, the normal approximation does not apply, so binomial probabilities are used to decide whether the test value is in the critical region or not.

For **example**, at the 5% level, the decision rules are:

- for a one-tailed (upper tail) test, reject H_0 if $P(X \geq \text{test value}) < 0.05$
- for a one-tailed (lower tail) test, reject H_0 if $P(X \leq \text{test value}) < 0.05$
- for a two-tailed test, reject H_0 if $P(X \leq \text{test value}) < 0.025$ or $P(X \geq \text{test value}) < 0.025$

Example

When a die was thrown 10 times, a six occurred 4 times. Is this evidence, at the 10% level of significance, that the die is biased in favour of sixes?

This is a small sample, so the normal approximation cannot be used.

Let the probability of obtaining a six be p.
If X is the number of sixes in 10 throws, then $X \sim B(10, p)$.

This is a one-tailed (upper tail) test.

$H_0: p = \frac{1}{6}$ (the die is fair), $H_1: p > \frac{1}{6}$ (the die is biased in favour of sixes).

If the null hypothesis is true, then $X \sim B(10, \frac{1}{6})$.
Reject H_0 if the test value of $x = 4$ lies in the critical region (the upper tail 10%)
i.e. reject H_0 if $P(X \geq 4) < 0.1$

It may be possible to calculate the probability using cumulative binomial tables. Check in your examination booklet.

$P(X \geq 4) = 1 - P(X < 4)$
$= 1 - ((\frac{5}{6})^{10} + {}^{10}C_1(\frac{5}{6})^9(\frac{1}{6}) + {}^{10}C_2(\frac{5}{6})^8(\frac{1}{6})^2 + {}^{10}C_3(\frac{5}{6})^7(\frac{1}{6})^3)$
$= 1 - 0.930 \ldots = 0.070$ (2 s.f.)

Since $P(X \geq 4) < 0.1$, H_0 is rejected.

There is evidence at the 10% level, that the die is biased in favour of sixes.

Test for a Poisson mean

The test statistic is X, the number of occurrences, where $X \sim Po(\lambda)$.

When λ is **small**, the test is similar to the small sample binomial test.

You will not be required to carry out this test.

When λ is **large**, $X \sim N(\lambda, \lambda)$ approximately and the test is similar to the large sample binomial test. A continuity correction is needed.

Probability and Statistics 2

Example

The office manager claims that the average number of telephone calls per minute received by her office switchboard is 4.5. Her supervisor maintains that it is less than 4.5. In a randomly selected minute, 2 calls were received. A hypothesis test is conducted at the 5% level. Whose claim is upheld?

Let X be the number of calls per minute. Assuming that calls occur randomly and independently, $X \sim \text{Po}(\lambda)$.

> This is a one-tailed (lower tail) test.

$H_0: \lambda = 4.5$, $H_1: \lambda < 4.5$. If the null hypothesis is true, $X \sim \text{Po}(4.5)$.

The test value is $x = 2$, so, at the 5% level, reject H_0 if $P(X \leq 2) < 0.05$

> Since λ is small, the method is similar to the small sample binomial test.

$$P(X \leq 2) = e^{-4.5} + 4.5e^{-4.5} + \frac{4.5^2}{2!}e^{-4.5} = e^{-4.5}\left(1 + 4.5 + \frac{4.5^2}{2!}\right) = 0.173\ldots$$

Since $P(X \leq 2) > 0.05$, the test value does not lie in the critical region and H_0 is not rejected. The office manager's claim that the mean is 4.5 is upheld.

Progress check

1. The variable X is modelled by a normal distribution with mean μ and standard deviation 2. The mean of a random sample of 10 items from X is 20.9. Test, at the 5% level, the hypotheses $H_0: \mu = 20$, $H_1: \mu > 20$.

2. The heights of a particular plant have mean μ and variance σ^2 (both unknown). The heights H, in centimetres, of a random sample of 100 plants are summarised as follows: $\Sigma h = 2941$, $\Sigma h^2 = 88\,502$.
   (a) Find unbiased estimates for μ and σ.
   (b) Does the sample data support the claim that the mean height is less than 30 cm? Perform a hypothesis test at the 10% level and state your hypotheses.

3. A random observation, x, is taken from $X \sim \text{B}(n, p)$.
   Test, at the 5% level, the null hypothesis that $p = 0.4$, against the alternative hypothesis that $p < 0.4$
   (a) if $n = 10$ and $x = 1$, (b) if $n = 100$ and $x = 32$.

> Hint:
> In (a) use small sample binomial test.
> In (b) use large sample binomial test.

4. A single observation is taken from a Poisson distribution with mean λ and used to test the null hypothesis $\lambda = 7$ against the alternative hypothesis $\lambda \neq 7$. What is the conclusion if the hypothesis test is carried out at the 10% level and the observation is
   (a) 3   (b) 13?

> Cumulative Poisson tables are needed for (b).

1. $z = 1.423$, do not reject H_0.
2. (a) 29.41, 4.503 (b) $H_0: \mu = 30$, $H_1: \mu > 30$; $z = -1.310$, lies in critical region so reject H_0 and uphold claim.
3. (a) $P(X \leq 1) = 0.0464 < 0.05$, reject H_0, conclude $p < 0.4$
   (b) $z = -1.531$, $z > -1.645$, do not reject H_0, conclude that p could be 0.4
4. (a) $P(X \leq 3) > 0.05$, H_0 is not rejected. (b) $P(X \geq 13) < 0.05$, H_0 is rejected.

Probability and Statistics 2

Sample questions and model answers

1

The continuous random variable X has probability density function $f(x)$ given by

$$f(x) = \begin{cases} kx(2-x) & 0 \leq x \leq 2 \\ 0 & \text{otherwise} \end{cases}$$

where k is a constant.

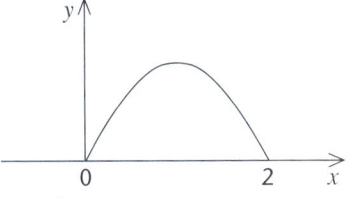

(a) Show that $k = \frac{3}{4}$ and find the exact value of $P(X > 1\frac{1}{2})$.

(b) Find $E(X)$ and $E(X^2)$.

(c) Show that the standard deviation, σ, is 0.447 correct to three decimal places.

The total area under the curve is 1.

As k is a constant, it can be taken outside the integration.

(a) $\int_0^2 f(x)dx = 1$

$\Rightarrow k\int_0^2 (2x - x^2)dx = 1$

$\Rightarrow k\left[x^2 - \frac{x^3}{3}\right]_0^2 = 1$

$\Rightarrow k(4 - \frac{8}{3} - 0) = 1$

$\Rightarrow k = \frac{3}{4}$

$f(x) = \frac{3}{4}(2x - x^2)$

$P(X > 1\frac{1}{2}) = \int_{1\frac{1}{2}}^2 f(x)dx$

$= \frac{3}{4}\int_{1\frac{1}{2}}^2 (2x - x^2)dx$

Substitute the limits carefully, working in fractions to obtain the exact answer.

$= \frac{3}{4}\left[x^2 - \frac{x^3}{3}\right]_{1\frac{1}{2}}^2$

$= \frac{3}{4}(4 - \frac{8}{3} - (\frac{9}{4} - \frac{9}{8}))$

$= \frac{5}{32}$

If you do not spot that $f(x)$ is symmetrical about $x = 1$, find $E(X)$ by integration.

(b) By symmetry, $E(X) = 1$

$E(X^2) = \int_0^2 x^2 f(x)dx$

$= \frac{3}{4}\int_0^2 (2x^3 - x^4)dx$

$= \frac{3}{4}\left[\frac{2x^4}{4} - \frac{x^5}{5}\right]_0^2$

$= \frac{3}{4}(8 - \frac{32}{5})$

$= 1.2$

(c) $Var(X) = E(X^2) - [E(X)]^2$

$= 1.2 - 1$

$= 0.2$

$\sigma = \sqrt{0.2} = 0.447$ (3 d.p.)

$[E(X)]^2$ is often written $E^2(X)$.

Probability and Statistics 2

Sample questions and model answers (continued)

2

In an advanced level examination taken by a large number of candidates, the marks were distributed normally with mean mark 68.7 and standard deviation 5.4.

A random sample of 100 scripts is taken and their mean mark denoted by $\bar{X}$.

(a) State the distribution of $\bar{X}$.

(b) Find the probability that the mean mark of the 100 scripts is between 68 and 70, giving your answer correct to 2 significant figures.

Let X be the examination mark, so $X \sim N(68.7, 5.4^2)$

(a) For samples of size 100,

$$\bar{X} \sim N\left(68.7, \frac{5.4^2}{100}\right), \text{ i.e. } \bar{X} \sim N(68.7, 0.2916)$$

(b) $P(68 < \bar{X} < 70) = P\left(\dfrac{68 - 68.7}{\sqrt{0.2916}} < Z < \dfrac{70 - 68.7}{\sqrt{0.2916}}\right)$

$= P(-1.296 < Z < 2.407)$

$= \Phi(1.296) + \Phi(2.407) - 1$

$= 0.9026 + 0.9919 - 1 = 0.89$ (2 s.f.)

> If the degree of approximation is not specified in the question or your examination rubric, then approximate sensibly.

Probability and Statistics 2

Sample questions and model answers (continued)

3

The lengths of a population of snakes are normally distributed with standard deviation 8 cm and unknown mean μ cm.

A random sample of 10 snakes is taken and their mean length calculated.

When a significance test, at the 10% level, is performed, the hypothesis that μ is 37.5 is rejected in favour of the hypothesis that it is greater than 37.5 cm.

(a) What can be said about the value of the sample mean?
(b) Explain briefly what is meant, in the context of this question, by a Type I error, and state the probability of making a Type I error.
(c) Find the probability of making a Type II error when $\mu = 41.1$

(a) If X is the length, in centimetres, then $X \sim N(\mu, 8^2)$.

$H_0: \mu = 37.5$
$H_1: \mu > 37.5$

For the significance test, the sampling distribution of means is considered.

If $\mu = 37.5$, then $\bar{X} \sim N\left(37.5, \dfrac{8^2}{10}\right)$.

From standard normal tables, $\Phi(z) = 0.9 \Rightarrow z = 1.282$

The critical value for a one-tailed (upper tail) at the 10% level is 1.282

H_0 is rejected if z is in the upper tail 10% of the distribution.

Since H_0 is rejected, $z > 1.282$,

Test statistic is $Z = \dfrac{\bar{X} - \mu}{\sigma/\sqrt{n}}$.

i.e. $\dfrac{\bar{x} - 37.5}{8/\sqrt{10}} > 1.282$

$\bar{x} > 40.74$

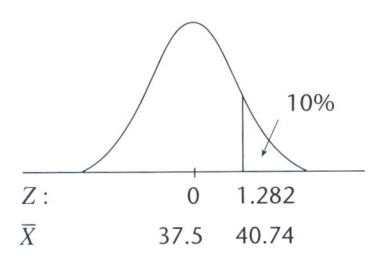

(b) A Type I error is made when the null hypothesis, $\mu = 37.5$, is rejected, when the mean population length is in fact 37.5

P(Type I error) = 10%

The probability of making a Type I error is the same as the level of significance of the test.

(c) $H_0: \mu = 37.5$
$H_1: \mu = 41.1$

P(Type II error) = P(H_0 is accepted | H_1 is true)

From (a), H_0 is accepted if $\bar{x} < 40.74$

If H_1 is true, $\bar{X} \sim N\left(41.1, \dfrac{8^2}{10}\right)$

P(Type II error) = $P\left(\bar{X} < 40.74 | \bar{X} \sim N\left(41.1, \dfrac{8^2}{10}\right)\right)$

$= P\left(Z < \dfrac{40.74 - 41.1}{8/\sqrt{10}}\right)$

$= P(Z < -0.142)$

$= 1 - 0.5565$

$= 0.44$ (2 s.f.)

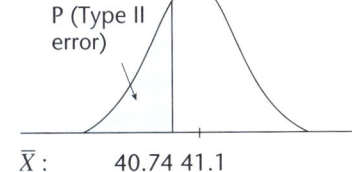

81

Probability and Statistics 2

Practice examination questions

1 The variable X is normally distributed with mean μ and variance 16.
The null hypothesis $\mu = 20$ is to be tested against the alternative hypothesis $\mu \neq 20$.

The mean of a random sample of 10 observations from X is 17.2.

Test at the 5% level whether this provides evidence that μ is not 20.

2 The probability that a person suffers from a particular health condition is 0.01.

(a) Using a suitable approximation, find the probability that in a randomly chosen sample of 80 people, fewer than 3 suffer from the health condition.

(b) Find the minimum sample size in order that the probability of including at least one with the health condition is greater than 95%.

3 The continuous random variable X has probability density function $y = k$, where k is a constant, for $2 \leq x \leq 12$ and $f(x) = 0$ otherwise.

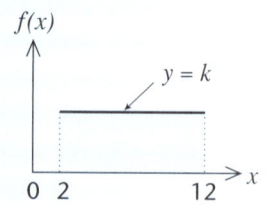

(a) Find the value of k.

(b) Calculate the probability that X lies within one standard deviation of the mean, giving your answer correct to 2 decimal places.

4 The volume, X litres, of paint in a tub may be modelled by a normal distribution with mean 3.5 and standard deviation σ.

(a) Assuming that $\sigma = 0.8$, find the value of $P(X < 3.3)$.
(b) Find the value of σ so that 95% of tubs contain more than 3.25 litres of paint.

5 The random variable X has a binomial distribution with $n = 10$ and $p = 0.4$.
The mean of 60 random observations of X is $\bar{X}$.

Find $P(\bar{X} < 3.5)$, explaining what part the central limit theorem has played in your answer.

Practice examination questions (continued)

6 A beetle infestation is discovered in the trees in a small copse. If more than 35% of the trees are infected it is impractical to treat the trees with chemicals and they will have to be felled.

BINOMIAL TEST
– small sample

(a) The representative from the ministry checked a random sample of 10 trees and found that 6 were infected. Stating any assumptions, use a hypothesis test, at the 10% significance level, to decide whether the infestation should be treated with chemicals or the trees in the copse felled.

BINOMIAL TEST
– large sample

(b) The estate manager checked a random sample of 30 trees and found that 14 were infected. Using a suitable approximation, carry out a significance test, again at the 10% level. On the basis of this sample, how should the infestation be dealt with?

7 The continuous random variable X has probability function $f(x)$ where $f(x) = k(x - x^3)$, $0 \leqslant x \leqslant 1$ and $f(x) = 0$ otherwise.

(a) Show that $k = 4$.
(b) Find $E(X)$.
(c) Find $P(X < 0.5)$.
(d) Find $Var(X)$.

8 A village newsagent finds that the demand for the local newspaper on a Monday follows a Poisson distribution with mean 16.

(a) Find the probability that there are fewer than 16 requests for the newspaper on a Monday.
(b) How many copies of the newspaper should there be in stock so that the demand is not satisfied on fewer than 5% of Mondays?

Chapter 3
Mechanics 2

The following topics are covered in this chapter:

- Projectiles
- Equilibrium of a rigid body
- Centre of mass
- Collisions and impulse
- Uniform circular motion
- Work, energy and power

3.1 Projectiles

After studying this section you should be able to:

- model the motion of a projectile moving under constant acceleration
- understand the limitations of the model
- solve problems by considering, separately, the vertical and horizontal components of velocity
- derive and apply the trajectory formula

LEARNING SUMMARY

Projectiles

In the standard model for projectiles, air resistance is ignored and the acceleration due to gravity g m s^{-2} is taken to be constant. According to this model:

- The only force acting on a projectile, once in flight, is its weight.
- The horizontal component of velocity is constant.
- The vertical component of velocity is subject to a constant acceleration g m s^{-2}.

KEY POINT

The approach used for many projectile problems is to resolve the initial velocity into horizontal and vertical components and to treat these two parts separately.

Example
A stone is thrown horizontally with a speed of 10 m s^{-1} from the top of a cliff. The cliff is 98 m high and the stone lands in the sea d m from its base. Take $g = 9.8$ m s^{-2} and show that $d = 20\sqrt{5}$.

You can assume that a cliff is vertical.

Vertically: $u = 0$
$a = 9.8$
$s = 98$.

Choosing downwards to be positive.

Using $s = ut + \frac{1}{2}at^2$
gives $98 = 4.9t^2$
$\Rightarrow t = \sqrt{20} = 2\sqrt{5}$.

Even a very simple diagram can help you to visualise the problem.

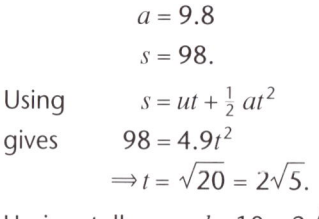

Horizontally: $d = 10 \times 2\sqrt{5}$
$\Rightarrow d = 20\sqrt{5}$.

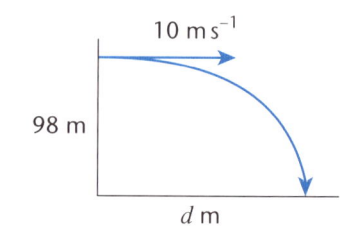

The initial velocity of the projectile may be at angle θ to the horizontal.

Example
A ball is struck with velocity 20 m s^{-1} at 40° to the horizontal from a point 1 m above the ground. Find the maximum height reached by the ball. Take $g = 9.8$ m s^{-2}.

Mechanics 2

Choosing upwards to be positive.

Vertically: $u = 20 \sin 40°$
$a = -9.8$

At max ht: $v = 0$.

Using $v = u + at$
$0 = 20 \sin 40° - 9.8t$
$\Rightarrow t = 1.3118 \ldots$

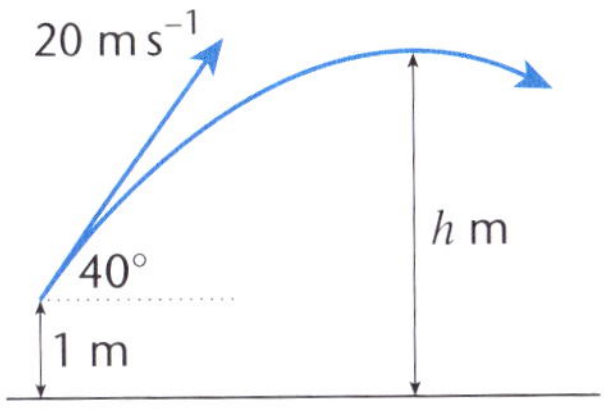

Using $s = ut + \frac{1}{2}at^2$.

At max ht $s = 20 \sin 40° \times 1.3118 - 4.9 \times 1.3118^2$
$= 8.432 \ldots$

The maximum height is 9.43 m to 2 d.p.

The model

In the usual **particle** model for projectiles, air resistance is ignored and the only force taken to act on the particle is its weight which is constant. According to this model:

Check the value of g used on your exam paper. Make sure that you use the given value or you may lose marks.

> **KEY POINT**
> - The horizontal component of velocity remains constant throughout the motion.
> - The vertical component of velocity is subject to a constant downward acceleration of g ms^{-2}.

The accuracy of the model, in predicting the motion of a projectile, depends on the extent to which the initial assumptions are satisfied.

For example, the model will provide accurate information about a projectile such as a stone or a dart over a short distance, but will give poor results for light objects that are affected more by air resistance.

Be careful about using formulae for time of flight, greatest height, range etc. without justification or, again, marks may be lost.

When using the model, a first step is usually to resolve the velocity of a projectile into horizontal and vertical components. The two parts are then treated separately.

A particle moving with speed v at angle θ to the horizontal has vertical component $v \sin \theta$ and horizontal component $v \cos \theta$.

The velocity is usually shown on a diagram as in Figure 1 but is used in equations as in Figure 2.

If you know the horizontal and vertical components of a velocity then you can combine them to find its magnitude (speed) and direction.

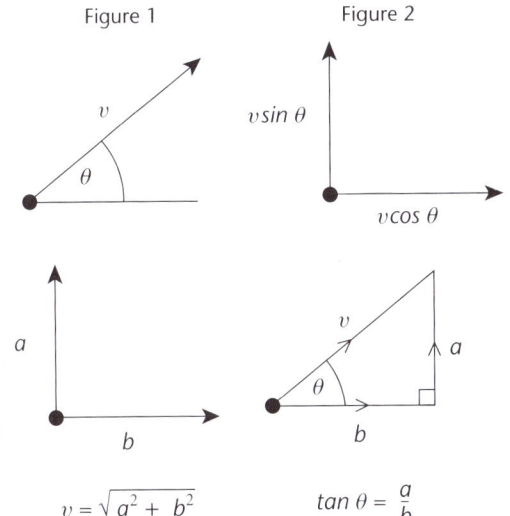

$v = \sqrt{a^2 + b^2}$ $\quad \tan \theta = \dfrac{a}{b}$

Example

A stone is thrown with speed 15 ms^{-1} at an angle of 60° above the horizontal. Find its speed and direction after 1 second. Take $g = 9.8$ ms^{-2}.

Upwards is taken to be positive so the acceleration is shown as negative.

Vertically: using $v = u + at$
$v_a = 15 \sin 60° - 9.8$
$= 3.19038 \ldots$

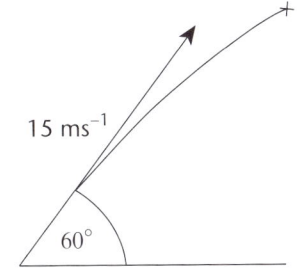

Horizontally: $v_b = 15 \cos 60°$
$= 7.5$
$v = \sqrt{7.5^2 + 3.19038^2} = 8.15$ to 3 s.f.

Mechanics 2

$$\tan\theta = \frac{3.19038}{7.5} \Rightarrow \theta = 23.0° \text{ to 1 d.p.}$$

After 1 second the stone has speed 8.15 ms^{-1} at 23° above the horizontal.

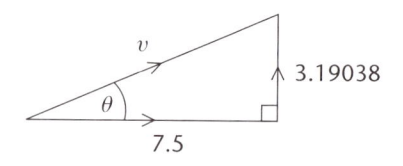

The trajectory formula

Taking the point of projection as the origin, the position of a projectile may be described in terms of (x, y) coordinates.

Horizontally: $x = v\cos\theta \times t \Rightarrow t = \dfrac{x}{v\cos\theta}$ (1)

Vertically: $y = v\sin\theta \times t - \tfrac{1}{2}gt^2$ (2)

Substituting for t in (2) gives: $y = v\sin\theta \times \dfrac{x}{v\cos\theta} - \dfrac{gx^2}{2v^2\cos^2\theta}$

so: $y = x\tan\theta - \dfrac{gx^2}{2v^2}(1 + \tan^2\theta)$

Note that: $\dfrac{1}{\cos^2\theta} \equiv \sec^2\theta \equiv (1 + \tan^2\theta)$.

> This is the Cartesian equation of the path of the projectile. It is known as the trajectory formula.

Example

A ball is kicked with speed 20 ms^{-1} at 30° above the horizontal towards a wall 5 m high. The wall is 15 m from the point where the ball is kicked. Take $g = 9.8$ ms^{-2}. Will the ball hit the wall or clear it? Consider the effect of any assumptions.

Using the formula: $y = x\tan\theta - \dfrac{gx^2}{2v^2}(1 + \tan^2\theta)$

$$y = 15\tan 30° - \dfrac{9.8 \times 15^2}{2 \times 20^2}(1 + \tan^2 30°)$$

$$= 4.99 \text{ m to 3 s.f.}$$

According to the model, the ball will hit the wall close to the top.
If the size of the ball is taken into account along with the effect of air resistance then it is clear that the ball cannot clear the wall.

Progress check

1. A small object is projected from ground level with speed 30 ms^{-1} at 40° above the horizontal. Take $g = 10$ ms^{-2}.
   (a) Find the speed and direction of the object after 1.5 seconds.
   (b) The object just clears a tree 25 m from the point of projection. How high is the tree?

1 (a) 23.4 ms^{-1}, 10.6° above horiz (b) 15.2 m

Mechanics 2

3.2 Equilibrium of a rigid body

After studying this section you should be able to:
- find the moment of a force
- understand the conditions for equilibrium of a rigid body under the action of coplanar forces
- solve problems involving coplanar forces

Moments

The **moment of a force** about a point is a measure of the turning effect of the force about the point.
It is found by multiplying the force by its perpendicular distance from the point.
The moment of a force acts either clockwise ↻ or anticlockwise ↺ about the point.

The units of force (N) are multiplied by the units of distance (m) to give N m.

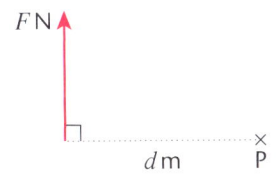

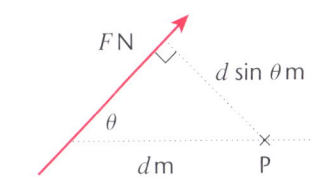

The moment of F about P is Fd N m ↻.

The moment of F about P is $Fd \sin \theta$ N m ↻.

When there is more than one force, the **resultant moment** about a point is found by adding the separate moments. One direction is taken to be positive and the other negative.

Example Find the resultant moment about O of the forces shown in the diagram.

Taking ↻ as positive:
The total moment is $8 \times 3 - 10 \times 2$ N m ↻
$= 4$ N m ↻.

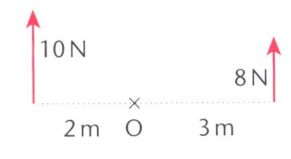

Equilibrium

For an object to be in **equilibrium**, the resultant force acting on it must be zero and the resultant moment must be zero.

In this context, a light rod is a rod is taken to have a mass that is so small that it can be ignored in any calculations.

Example
The diagram shows a light rod AB in equilibrium. Find the values of F and d.

Resolving vertically $\quad F - 30 - 20 = 0$
$\Rightarrow \quad F = 50$

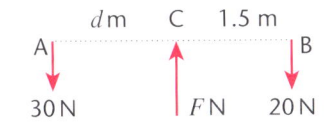

Taking moments about C (written as M(C)) will give an equation involving d.

M(C) gives $\quad 30d - 20 \times 1.5 = 0$
$\Rightarrow \quad 30d = 30$
$\Rightarrow \quad d = 1.$

87

Mechanics 2

Forces that act in the same plane are coplanar.

You may need to show that a rigid body is in equilibrium under a system of **coplanar** forces *or* to *use* the fact that a rigid body is in equilibrium to find something about the forces acting on it.

To show that a rigid body is in equilibrium under the action of coplanar forces you need to establish that:

- the vector sum of the forces is zero
- the sum of the moments about some point is zero.

You only need to establish this for one *point and you can choose any point that is convenient.*

If a body is in equilibrium under some coplanar forces then you know that:

- the vector sum of the forces will be zero
- the sum of the moments about any point you choose will be zero.

> **KEY POINT**
> The conditions for equilibrium can be used to produce equations. The equations can then be solved to find the unknown values in a problem.

Example

A uniform plank AB of length 5 m and mass 20 kg rests horizontally on two supports, one at each end. A mass of 10 kg is positioned on the plank 1 m from B. Find the reaction at each of the supports. Take $g = 9.8$ ms^{-2}.

The first step is to draw a clearly labelled diagram.

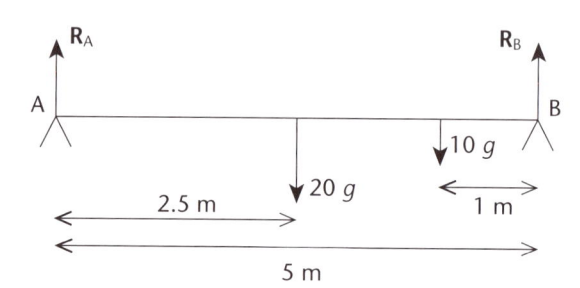

The plank is uniform so its weight acts at its mid-point.

The vector sum of the forces is zero so the total upward force must equal the total downward force.

Vertically: $\quad R_A + R_B = 20g + 10g$
$\qquad\qquad\qquad\quad = 30g$

M(A) gives: $\quad 5R_B = 20g \times 2.5 + 10g \times 4$
$\qquad\qquad\qquad = 90g$
$\qquad\qquad R_B = 18g$
$\qquad\qquad\quad\; = 176.4$ N (reaction at B)

Substitute for R_B: $\quad R_A + 18g = 30g$
$\qquad\qquad\qquad\qquad R_A = 12g$
$\qquad\qquad\qquad\qquad\quad\; = 117.6$ N (reaction at A).

Mechanics 2

Leaning ladders

The situation where a ladder leans against a wall, possibly supporting a load at some point along its length, provides a rich source of questions on equilibrium. It requires forces and moments to be considered and includes the theory of friction.

Example

A uniform ladder of mass m kg rests against a smooth vertical wall with its lower end on rough horizontal ground. A man of mass $4m$ kg has climbed three-quarters of the way up the ladder but cannot go any higher or it will slip. The ladder makes an angle of $60°$ with the horizontal.
Find the coefficient of friction between the ladder and the ground.

Start by drawing a clearly labelled diagram.

> The wall is smooth so there is no vertical force at the top of the ladder.

> The length of the ladder has been written as $4l$ to simplify taking moments.

> Friction acts at the bottom of the ladder to oppose the tendency for it to slip.

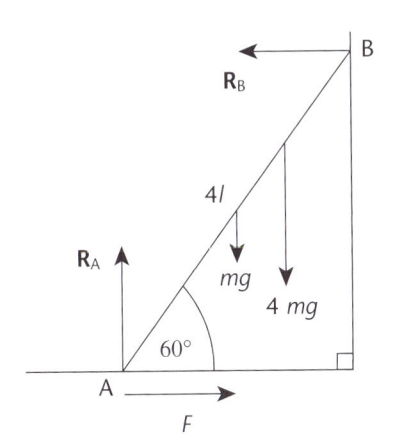

In this example, three equations may be formed by resolving the forces in two directions and taking moments.

The equations provide enough information to solve the problem.

Vertically: $\qquad R_A = mg + 4mg$ (1)
$\qquad\qquad\qquad = 5mg$

Horizontally: $\qquad F = R_B$ (2)

M(A): $\qquad R_B \times 4l \sin 60° = mg \times 2l \cos 60° + 4mg \times 3l \cos 60°$

giving: $\qquad R_B \times 2\sqrt{3}\,l = mgl + 6mgl$

so: $\qquad R_B = \dfrac{7mg}{2\sqrt{3}}$ (3)

From (2) and (3) $\qquad F = \dfrac{7mg}{2\sqrt{3}}$

Since the ladder is on the point of slipping, friction is limiting and the coefficient of friction is given by:

$$\mu = \dfrac{F}{R_A} = \dfrac{7}{10\sqrt{3}} = \dfrac{7\sqrt{3}}{30}$$

> It's a good idea to rationalise the denominator by multiplying the top and bottom of the fraction by $\sqrt{3}$.

Progress check

1. A uniform plank AB of length 3 m and mass 25 kg rests horizontally on two supports, one at each end. A mass of 15 kg is positioned on the plank 1 m from B. Find the reaction at each of the supports. Take $g = 9.8$ ms^{-2}.

2. A uniform ladder of length 5 m rests against a smooth wall with its base on rough horizontal ground. The coefficient of friction between the ladder and the ground is $\frac{2}{3}$ and the ladder is on the point of slipping. Show that the angle between the ladder and the ground is $\tan^{-1}(\frac{3}{4})$.

1 220.5 N (A), 171.5 N (B)

89

Mechanics 2

3.3 Centre of mass

After studying this section you should be able to:

- find the centre of mass of a system of particles in one and two dimensions
- find the centre of mass of a lamina by considering an equivalent system of particles
- use centre of mass to determine conditions for the equilibrium of a lamina

Centre of mass of a system of particles

In the diagram, m_1 and m_2 lie on a straight line through O. Their displacements from O are x_1 and x_2 respectively.

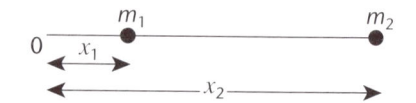

The position given by $\bar{x} = \dfrac{m_1 x_1 + m_2 x_2}{m_1 + m_2}$ is called the **centre of mass** of m_1 and m_2.

> If m_1 and m_2 are equal then this result gives the mid-point of the two masses.

For n separate masses $m_1, m_2, \ldots m_n$ the position of the centre of mass is given by

$$\bar{x} = \frac{m_1 x_1 + m_2 x_2 + \ldots m_n x_n}{m_1 + m_2 + \ldots m_n}$$

This is usually written as $\bar{x} = \dfrac{\sum_{i=1}^{n} m_i x_i}{\sum_{i=1}^{n} m_i}$

In two dimensions, using the usual Cartesian coordinates, the position of the centre of mass is at $(\bar{x}, \bar{y})$ where $\bar{x}$ and $\bar{y}$ are given by

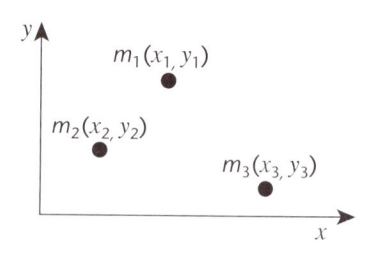

$$\bar{x} = \frac{\sum_{i=1}^{n} m_i x_i}{\sum_{i=1}^{n} m_i} \text{ as above, and } \bar{y} = \frac{\sum_{i=1}^{n} m_i y_i}{\sum_{i=1}^{n} m_i}$$

> The method is easily extended to deal with any number of masses.

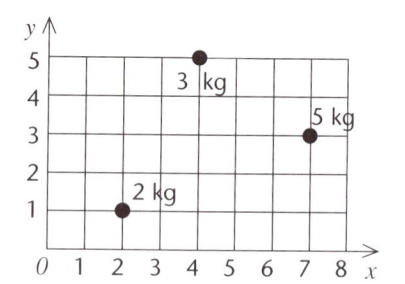

Example

Masses of 2 kg, 3 kg and 5 kg are positioned as shown in the diagram.
Find the coordinates of the centre of mass of the system.

$$\bar{x} = \frac{2 \times 2 + 3 \times 4 + 5 \times 7}{2 + 3 + 5} = \frac{51}{10} = 5.1$$

$$\bar{y} = \frac{2 \times 1 + 3 \times 5 + 5 \times 3}{2 + 3 + 5} = \frac{32}{10} = 3.2$$

The centre of mass has coordinates (5.1, 3.2)

Mechanics 2

Centre of mass of a lamina

A **lamina** is something that is thin and flat such as a sheet of metal. The thickness of a lamina is ignored and it is treated as a two-dimensional object.

You can find the centre of mass of a uniform lamina by dividing it into parts which you then represent as particles. Each part is usually rectangular or triangular and in order to represent it as a particle:

- the mass of the particle is represented by the area of the part
- the position of the particle is taken to be at the geometric centre of the part.

This representation of a lamina gives you the information you need to apply the formulae for its centre of mass.

Example

Find the coordinates of the centre of mass of this uniform lamina.

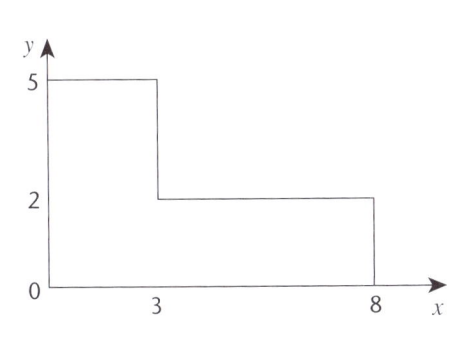

There are different ways to do this, but it doesn't matter which you choose.

The first step is to divide the lamina into rectangles.

Rectangle A has area 15 square units. Its centre is at (1.5, 2.5).

Rectangle B has area 10 square units. Its centre is at (5.5, 1).

This process is easily extended to deal with more complex composite figures.

$$\bar{x} = \frac{15 \times 1.5 + 10 \times 5.5}{15 + 10} = 3.1$$

$$\bar{y} = \frac{15 \times 2.5 + 10 \times 1}{15 + 10} = 1.9$$

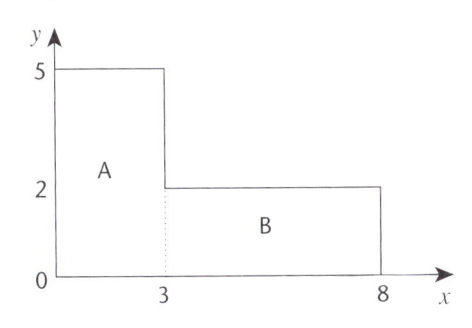

The centre of mass of the lamina has coordinates (3.1, 1.9).

Freely suspended lamina

Once you know where the centre of mass of a lamina is, you can use this to work out how it will move if it is freely suspended from a given point.

> **KEY POINT**
> When a lamina is freely suspended, it will move so that its centre of mass lies directly below the point of suspension.

Mechanics 2

Example

The diagram shows the lamina of the previous example hanging freely from one corner.

Find the size of angle θ between the direction of Oy and the horizontal.

The position of the centre of mass is shown at the point G.
Notice that θ also lies in a right-angled triangle in which two of the sides are known.

From the diagram: $\tan \theta = \dfrac{1.9}{3.1}$

giving $\theta = 31.5°$ to 1 d.p.

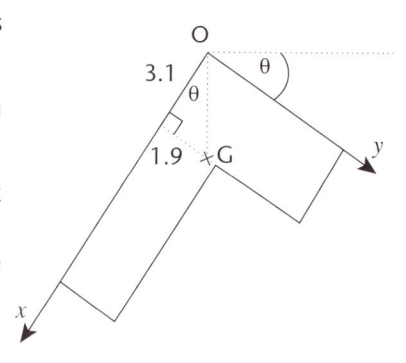

Progress check

1. Masses of 2 kg, 3 kg and 5 kg are positioned as shown in the diagram.
   (a) Find the coordinates of G, the centre of mass of the system.
   (b) Find the angle θ between OG and Ox.

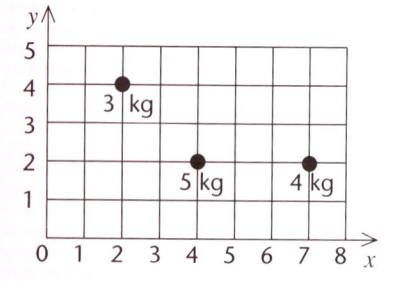

1 (a) G(4.5, 2.5) (b) 29.1°

Mechanics 2

3.4 Collisions and impulse

After studying this section you should be able to:

- apply the principle of conservation of momentum to direct impact
- apply Newton's experimental law using the coefficient of restitution
- calculate the loss of mechanical energy in a collision
- recall and use the definition of impulse as a change of momentum

Conservation of momentum

The **momentum** of an object is the product of its mass and its velocity, i.e. momentum = $m\mathbf{v}$. Momentum has direction and so it is a vector quantity.

Momentum is measured in Ns (Newton seconds).

It's important to use the correct units. Remember that speed is the magnitude of the velocity.

For example, an object of mass 5 kg moving with speed 6 ms^{-1} in a given direction has momentum $5 \times 6 = 30$ Ns in the direction of movement.

The **principle of conservation of momentum** states that, in a collision, the total momentum before impact = the total momentum after impact.

Key points from AS

- **Momentum**
  Revise AS page 67

Remember:

- always draw a diagram to represent the situation
- choose one direction to be positive (usually left to right $\rightarrow$)
- show any unknown velocities in the positive direction.

Example

Two particles A and B of masses 4 kg and 6 kg respectively move towards each other along the same straight line. Particle A is moving with speed 3 ms^{-1} and particle B is moving with speed 5 ms^{-1}. Given that the particles coalesce on impact, find their common velocity immediately afterwards.

When particles coalesce, they stick together and behave as a single particle.

The velocity of B is in the negative direction.

Before impact:

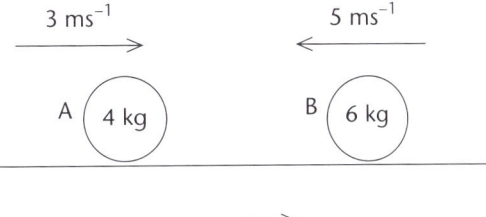

$\quad\quad$ 3 ms^{-1} $\rightarrow$ $\quad\quad$ 5 ms^{-1} $\leftarrow$

$\quad\quad$ A (4 kg) $\quad\quad$ B (6 kg)

After impact:

$\quad\quad$ v ms^{-1} $\rightarrow$

By the principle of conservation of momentum:

$$4 \times 3 - 6 \times 5 = 10v$$

giving: $\quad 10v = -18$

so: $\quad v = -1.8$

v is negative so the direction of the velocity is from right to left.

The particles move with velocity 1.8 ms^{-1} in the direction from B to A.

Mechanics 2

Newton's experimental law

Some problems require two unknown values to be found and in this situation a second equation is needed.

Newton found by experiment that, when two objects collide, the ratio of their speed of separation to their speed of approach has a fixed value. The symbol used to represent this ratio is e and its value is depends on what the objects are made of.

$$e = \frac{\text{speed of separation}}{\text{speed of approach}}$$

This is known as **Newton's experimental law** and is often used to provide a second equation.

This is often used in the form $e \times$ speed of approach = speed of separation.

The speed of separation is always less than or equal to the speed of approach so it follows that $0 \leq e \leq 1$. The ratio e is called the **coefficient of restitution**.

Example

A smooth sphere P moves towards a stationary smooth sphere Q in a straight line through their line of centres. P has mass 2 kg and speed 5 ms^{-1}. Q has mass 4 kg. Given that the coefficient of restitution between the spheres is 0.5, find:

(a) the speed of each sphere immediately after impact
(b) the loss of energy in the collision.

> In the exam, you may need to deal with more than one impact. See the sample exam questions page 102.

(a)

Before impact:

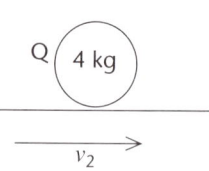

After impact:

Momentum is conserved:
$2 \times 5 + 4 \times 0 = 2v_1 + 4v_2$
So: $v_1 + 2v_2 = 5$ (1)

Newton's law: $0.5 \times 5 = v_2 - v_1$ (2)

Adding (1) and (2) gives: $3v_2 = 7.5 \Rightarrow v_2 = 2.5$
and: $v_1 = 0$

Immediately after the impact, P is brought to rest and Q moves with speed 2.5 ms^{-1}.

> Kinetic energy = $\frac{1}{2}mv^2$. Even though momentum is conserved, energy is lost.

(b) Loss of energy = KE before impact − KE after impact
= $0.5 \times 2 \times 5^2 - 0.5 \times 4 \times 2.5^2$
= 12.5 J

Mechanics 2

Impulse

The **impulse** of a constant force acting over a given time is given by the product of force and time. This may be written as $I = Ft$ where the impulse is I N s, the force is F N and the time is t s.

It follows that: Impulse = change in momentum

so $Ft = mv - mu$.

Example

A particle of mass 4 kg, initially at rest, is acted upon by a force of 3 N for 10 s. What is the speed of the particle at the end of this time?

Using $Ft = mv - mu$

gives $3 \times 10 = 4v - 0$

$\Rightarrow \quad v = 7.5$

The speed of the particle is 7.5 m s^{-1}.

When two particles collide, the contact force between them may only last a very short time and is unlikely to be constant. However, the value of F in the equation $Ft = mv - mu$ may be used to represent the average value of this force.

Example

A ball of mass 0.5 kg strikes a hard floor with speed 2 m s^{-1} and rebounds with speed 1.5 m s^{-1}.
Given that the ball is in contact with the floor for 0.05 s find the average value of the contact force.

Taking upwards as the positive direction:

$u = -2$, $v = 1.5$, $m = 0.5$ and $t = 0.05$

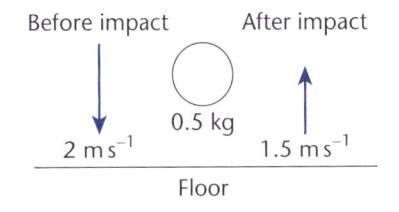

Using $Ft = mv - mu$

gives $0.05F = 0.5 \times 1.5 - 0.5 \times (-2)$

$\Rightarrow \quad 0.05F = 1.75$

$\Rightarrow \quad F = 35$

The average value of the contact force is 35 N.

Impulse = change in momentum

> The total momentum of the system is conserved.

When two particles collide, each receives an impulse from the other of equal size but opposite sign. In this way, the total change in momentum is zero. We would expect this to happen because the total momentum of the system is conserved in a collision.

Mechanics 2

Example

A ball of mass 0.5 kg moving with velocity $6\mathbf{i} + 2\mathbf{j}$ ms^{-1} is kicked and receives an impulse of $2\mathbf{i} + 11\mathbf{j}$ Ns. Find the velocity of the ball immediately after it is kicked.

Momentum immediately after kick = initial momentum + impulse
$$= 0.5(6\mathbf{i} + 2\mathbf{j}) + 2\mathbf{i} + 11\mathbf{j} = 5\mathbf{i} + 12\mathbf{j}$$
$$= 0.5\mathbf{v} \text{ (where } \mathbf{v} \text{ is the velocity)}$$

This gives: $\mathbf{v} = 10\mathbf{i} + 24\mathbf{j}$ ms^{-1}

Progress check

1. Two particles A and B move towards each other with speeds of 6 ms^{-1} and 2 ms^{-1} respectively. A has mass 3 kg and B has mass 1 kg. The coefficient of restitution between the particles is 0.6. Find the speed of each particle after the collision.

2. A particle of mass 8 kg moving with speed 2 m s^{-1} hits a stationary particle of mass 2 kg.
   After the impact both particles move in the same direction with speed v m s^{-1}.
   (a) Find the value of v.
   (b) Find the impulse given to the stationary particle.

1. A 2.8 ms^{-1}, B 7.6 ms^{-1}.
2. (a) $v = 1.6$ (b) Impulse = 3.2 N s.

Mechanics 2

3.5 Uniform circular motion

After studying this section you should be able to:

- understand angular speed and be able to apply the formula $v = r\omega$
- understand that a particle moving in a circular path with constant speed has an acceleration towards the centre of the circle
- use the formula $\omega^2 r$, or its equivalent form $\dfrac{v^2}{r}$, to represent the magnitude of the acceleration towards the centre of the circle in solving problems

LEARNING SUMMARY

Angular speed

Suppose that a particle P moves with constant speed in a circular path with centre O.

The direction of OP changes with time. The angle θ, through which OP turns in some given time, is usually measured in radians.

The **angular speed** of P is the rate of change of θ with respect to time. Using ω to represent angular speed in radians per second gives:

$$\omega = \frac{\theta}{t}$$

One advantage of using radians is that the relationship between angular speed ω, and linear speed v, is easily expressed as $v = r\omega$.

For example, a particle moving in a circular path of radius 2 m with angular speed 10 radians/sec has a straight line speed of $2 \times 10 = 20$ ms^{-1}.

Radial acceleration

Notice that while the speed of the particle may be constant, its *direction* changes as it moves around the circle. This means that the *velocity* is *not constant* and so the particle must be *accelerating*.

This acceleration is always directed towards the centre of the circle and is known as **radial acceleration** or sometimes **centripetal acceleration**.

The magnitude of this acceleration is given by $a = \omega^2 r$.

Since $v = r\omega$, it follows that $a = \dfrac{v^2}{r}$.

This means that there is a choice of which formula to use. In practice, you should use the one that best fits the available information.

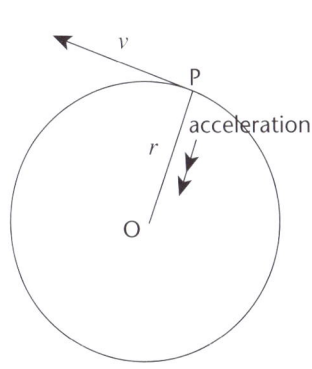

Example

A particle moves in a circular path of radius 3 m with angular speed 5 radians/sec. Find the magnitude of its radial acceleration.

Using $a = \omega^2 r$: the radial acceleration is $5^2 \times 3 = 75$ ms^{-2}.

> Since we are given the angular speed, the formula $a = \omega^2 r$ is the simpler one to use here.

97

Mechanics 2

Circular motion and force

The radial acceleration of an object moving in a circular path must be produced as the result of a *force acting towards the centre of the circle*. This force is sometimes called the **centripetal force**.

Using $F = ma$, the force needed to keep an object of mass m kg moving in a circular path is $m\omega^2 r$. An alternative form of this is $\dfrac{mv^2}{r}$.

In applying this theory to solving problems, the centripetal force may take various forms such as a tension in a string, friction or the gravitational attraction of a planet.

Example

One end of a string of length 80 cm is attached to a point O on a smooth horizontal table. The other end is attached to a mass of 3 kg moving with speed 4 ms^{-1} and the string is taut. Find the tension in the string.

> It is important to write the length of the string in metres.

Using $T = \dfrac{mv^2}{r}$ gives $T = \dfrac{3 \times 4^2}{0.8} = 60$

The tension in the string is 60 N.

The conical pendulum

One particular situation involving circular motion is described as the **conical pendulum**.

This is where an object moves in a horizontal circle while suspended by a string attached to a point vertically above the centre of the circle.

- The vertical component of the tension in the string supports the weight of the object.
- The horizontal component of the tension in the string provides the centripetal force.

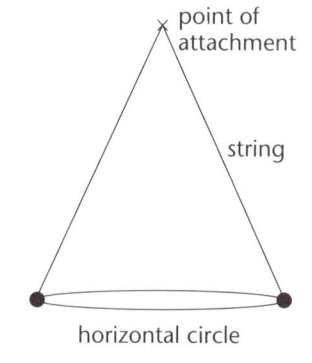

Example

A light inextensible string is fixed at one end to a point O. The other end is attached to a particle of mass 2 kg which moves in a horizontal circle of radius 30 cm with angular speed 4 radians per second. Find the angle of inclination of the string to the vertical. Take $g = 9.8$ ms^{-2}.

Vertically: $T \cos \theta = 19.6$ (1)

Horizontally: $T \sin \theta = 2 \times 4^2 \times 0.3 = 9.6$ (2)

(2) ÷ (1) gives: $\tan \theta = \dfrac{9.6}{19.6} \Rightarrow \theta = 26.1°$

The string is inclined at 26.1° to the vertical.

Progress check

1. One end of a string of length 60 cm is attached to a point O on a smooth horizontal table. The other end is attached to a mass of 5 kg moving with speed 3 ms^{-1} and the string is taut. Find the tension in the string.

1. 75 N

Mechanics 2

3.6 Work, energy and power

LEARNING SUMMARY

After studying this section you should be able to:
- understand kinetic and potential energy and the work energy principle
- understand the principle of conservation of mechanical energy
- understand the definition of power
- solve problems involving work, energy and power

Work and energy

In everyday language, **energy** is needed to do **work**. This idea is given a formal interpretation within mechanics.

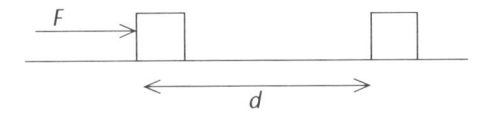

The work done, W, by a *constant* force, F, in moving an object through a distance, d, along its line of action is given by $W = Fd$. The unit of work is the **joule** (J).

Energy is also measured in joules and the energy used in moving the object is equal to the work done. In a way, work and energy are like two sides of the same coin.

Energy may take different forms such as heat, sound, electrical, chemical and mechanical. An important property of energy is that it may change from one form to another. For your mechanics module, the focus of attention is on mechanical energy.

There are two forms of mechanical energy: **kinetic energy (KE)** and **potential energy (PE)**

The kinetic energy of an object is the energy it has due to its *motion*.

Its value depends on the mass of the object and its speed. $KE = \frac{1}{2}mv^2$.

For example, a mass of 4 kg moving with speed 5 ms^{-1} has $KE = \frac{1}{2} \times 4 \times 5^2 = 50$ J

The potential energy of an object is the energy it has due to its *position*.

The **gravitational potential energy** of an object, relative to some level, is equal to the work done against gravity in raising the object from that level to its current position.

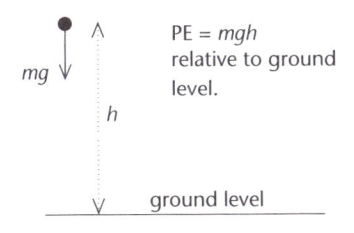

$PE = mgh$ relative to ground level.

KEY POINT

The total mechanical energy of an object is the sum of its KE and PE.

The **principle of conservation of mechanical energy** states that the total mechanical energy of a system remains constant, provided that it is not acted on by any external force, other than gravity.

The principle of conservation of energy applies, for example, to the motion of a projectile. Any change in the PE of the projectile corresponds to an equal but opposite change in its KE.

99

Mechanics 2

Power

Power is the rate of doing work. It is measured in watts (W) where $1\text{ W} = 1\text{ Js}^{-1}$.

If the point of application of a force F moves with speed v in the *direction of the force* then the power P is given by $P = Fv$.

Example

Find the power generated by a car engine when the car travels at a constant speed of 72 kmh^{-1} on horizontal ground against resistance forces of 2500 N.

$$72\text{ kmh}^{-1} = \frac{72 \times 1000}{60 \times 60}\text{ ms}^{-1} = 20\text{ ms}^{-1}.$$

Power $= 2500 \times 20\text{ W} = 50\,000\text{ W}$
      $= 50\text{ kW}$

> Since the speed is constant, the force produced by the engine must match the resistance force.

Exam questions may involve resisted motion on an inclined plane.

Example

A car of mass 1000 kg travels up a hill inclined at angle θ to the horizontal, where $\sin\theta = \frac{1}{20}$. The non-gravitational resistance to motion is 2000 N and the power output from the engine is 60 kW. Find the acceleration of the car when it is travelling at 10 ms^{-1}. Take $g = 10\text{ ms}^{-2}$.

> A clearly labelled diagram is a must.

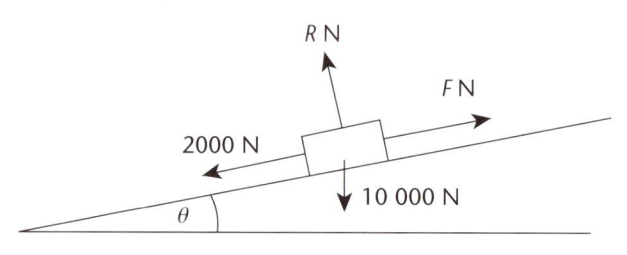

[Diagram: block on incline at angle θ; forces labelled R N (normal), F N (up slope), 2000 N (down slope), 10 000 N (weight down).]

> In this formula, F represents the *resultant force* up the hill.

Using $P = Fv$: $\quad 60\,000 = F \times 10 \Rightarrow F = 6000$

Using $F = ma$: $\quad 6000 - 2000 - 10\,000 \times \frac{1}{20} = 1000a$

This gives: $\quad 3500 = 1000a \Rightarrow a = 3.5$

The acceleration of the car up the hill is 3.5 ms^{-2}.

Progress check

1. A ball is kicked from ground level with speed 15 ms^{-1} and just clears a wall of height 2 m. The speed of the ball as it passes over the wall is $v\text{ ms}^{-1}$.
   (a) Assume that the ball has mass m kg and use the conservation of mechanical energy principle to write an equation.
   (b) Solve the equation to find v.

2. A car of mass 800 kg travels up a hill inclined at angle θ to the horizontal, where $\sin\theta = \frac{1}{20}$. The non-gravitational resistance to motion is 1600 N and the power output from the engine is 45 kW. Find the acceleration of the car when it is travelling at 15 ms^{-1}. Take $g = 10\text{ ms}^{-2}$.

1 (a) $\frac{1}{2}m \times 15^2 = m \times 9.8 \times 2 + \frac{1}{2}mv^2$
  (b) 13.6 ms^{-1}
2 1.25 ms^{-2}

Mechanics 2

Sample questions and model answers

1

A ball of mass 250 g is released from rest 2 m above the ground. The coefficient of restitution between the ball and the ground is 0.7. Take $g = 9.8$ ms^{-2}.

(a) Find the height reached by the ball as it rebounds from the ground.
(b) Find the impulse exerted on the ball by the ground.

(a) Using $v^2 = u^2 + 2as$
gives $v^2 = 2 \times 9.8 \times 2$
$= 39.2$
so $v = \sqrt{39.2}$

There is no need to work out $\sqrt{39.2}$ at this stage.

The ball strikes the floor with speed $\sqrt{39.2}$ ms^{-1}.

Using Newton's law, the ball rebounds with speed $\sqrt{39.2} \times 0.7$ ms^{-1}

Some simple statements help to make your method clear.

Using $v^2 = u^2 + 2as$
gives $0 = 39.2 \times 0.49 - 2 \times 9.8 \times h$

so $h = \dfrac{39.2 \times 0.49}{2 \times 9.8} = 0.98$

The ball reaches a height of 0.98 m when it rebounds from the ground.

(b) Taking upwards as positive:

Momentum of ball immediately before it hits the ground
$= -0.25 \times \sqrt{39.2}$ Ns

Write the mass in kg.

Momentum of ball immediately after it hits the ground
$= 0.25 \times \sqrt{39.2} \times 0.7$ Ns

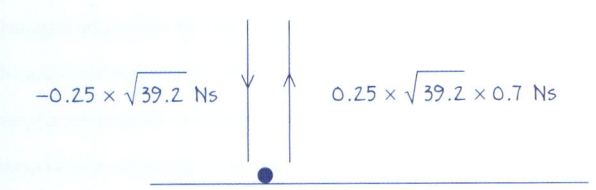

$-0.25 \times \sqrt{39.2}$ Ns $0.25 \times \sqrt{39.2} \times 0.7$ Ns

Impulse = change in momentum
$= 0.25 \times \sqrt{39.2} \times 0.7 - (-0.25 \times \sqrt{39.2})$
$= 2.66$ Ns to 3 s.f.

The impulse exerted on the ball by the ground is 2.66 Ns.

Mechanics 2

Sample questions and model answers (continued)

2

A particle P of mass $3m$ is moving with speed $2u$ when it strikes a particle Q of mass m which is at rest. The coefficient of restitution between the particles is e.

(a) Show that the speed of P after the collision is $\frac{v}{2}(3-e)$ and find the speed of Q.

(b) Q subsequently strikes a wall perpendicular to its direction of motion and rebounds so that a second collision with P occurs. Given that the coefficient of restitution between Q and the wall is $\frac{1}{3}$ find the speed of P after the second collision.

(c) Show that P moves in the same direction after the second collision.

(a) Before impact

$2u \rightarrow$

P (3m)   Q (m)

After impact

$v_P \rightarrow$ $v_Q \rightarrow$

The approach is exactly the same as when numerical values of the speeds are given.

Conservation of momentum gives: $6mu = 3mv_P + mv_Q$

which simplifies to $\quad 6u = 3v_P + v_Q \quad (1)$

Newton's experimental law gives: $2ue = v_Q - v_P \quad (2)$

(1) − (2) gives: $\quad 6u - 2ue = 4v_P$

giving $\quad v_P = \dfrac{u}{2}(3-e)$ as required

This is the result given in the question.

and from (2) $\quad v_Q = 2ue + v_P = 2ue + \dfrac{u}{2}(3-e)$

$= \dfrac{u}{2}(4e + 3 - e) = \dfrac{3u}{2}(e+1)$

The speed of Q after the collision is $\dfrac{3u}{2}(e+1)$.

Mechanics 2

Sample questions and model answers (continued)

(b) The speed of Q after hitting the wall is $\dfrac{1}{3} \times \dfrac{3u}{2}(e+1) = \dfrac{u}{2}(e+1)$ in the opposite direction.

Repeat the process with the new speeds. Draw a diagram and set your working out clearly.

Before impact:

$\dfrac{u}{2}(3-e) \longrightarrow \qquad \longleftarrow \dfrac{u}{2}(e+1)$

P (3m)   Q (m)

After impact:

$\longrightarrow w_P \qquad \longrightarrow w_Q$

Conservation of momentum gives: $\dfrac{3mu}{2}(3-e) - \dfrac{mu}{2}(e+1) = 3mw_P + mw_Q$

which simplifies to $\qquad 4u - 2ue = 3w_P + w_Q \qquad (3)$

Newton's experimental law gives: $\left(\dfrac{u}{2}(3-e) + \dfrac{u}{2}(e+1)\right)e = w_Q - w_P$

which simplifies to $\qquad 2u = w_Q - w_P \qquad (4)$

(3) − (4) gives $\qquad 2u - 2ue = 4w_P$

and $\qquad w_P = \dfrac{u}{2}(1-e)$

There is no need to find w_Q.

so the speed of P immediately after the second collision is $w_P = \dfrac{u}{2}(1-e)$.

(c) $1 - e \geq 0$ since $0 \leq e \leq 1$ so the direction of movement of P is not changed.

Mechanics 2

Sample questions and model answers (continued)

3

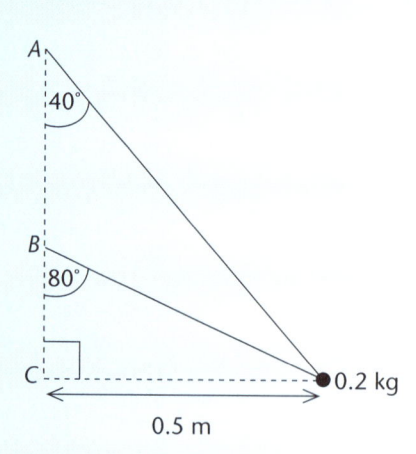

The diagram shows a particle of mass 0.2 kg threaded onto a smooth string fixed at its end-points A and B. The points A, B and C lie in a vertical line and the particle travels in a horizontal circle of radius 0.5 m with centre C.

(a) Calculate the tension in the string.

(b) Calculate the speed of the particle.

(a) Since the string is smooth, the tension in both parts of the string is the same.

The diagram shows the forces acting on the particle.

There is no acceleration vertically.
Resolving vertically $T\cos 40° + T\cos 80° = 0.2 \times 9.8$

$$T = \frac{0.2 \times 9.8}{0.93969} = 2.0857\ldots$$

The tension in the string is 2.09 N (3 s.f.)

(b) Let the speed of the particle be v ms^{-1}.

Centripetal force $= \frac{mv^2}{r}$

Resolving horizontally   Centripetal force $= T\sin 40° + T\sin 80°$

$$= 3.3948\ldots = \frac{0.2v^2}{0.5}$$

$$v = \sqrt{\frac{3.3948}{0.4}}$$

$$= 2.913\ldots$$

The speed of the particle is 2.91 ms^{-1} (3 s.f.)

Mechanics 2

Practice examination questions

1 A ball is struck with speed 25 ms⁻¹ from a point 50 cm above the ground. When it has travelled 20 m horizontally, the ball just clears a fence that is 3 m high.

(a) Show that a possible angle of projection of the ball is 16.5°.

(b) Find the distance d m, between the fence and the point where the ball strikes the ground for this angle of projection.

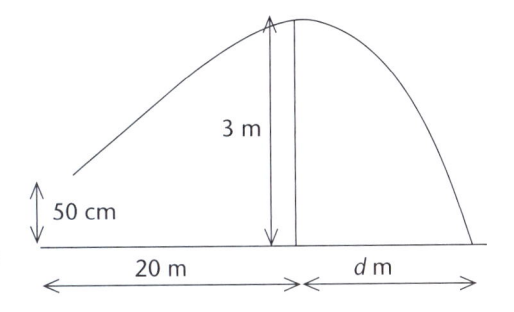

2 The diagram shows a uniform lamina ABCDEFGH.

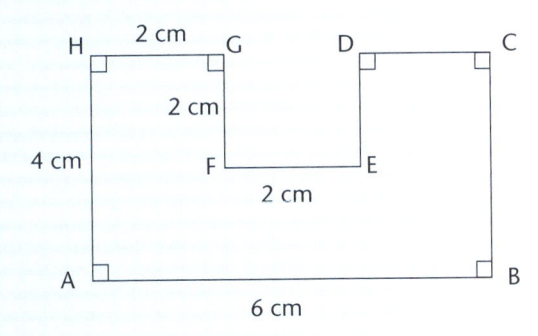

(a) Write down the distance of the centre of mass of the lamina from AH.

(b) Find the distance of the centre of mass of the lamina from AB.

(c) The lamina is now suspended freely from A. Find the angle that AB makes with the horizontal in its equilibrium position.

3

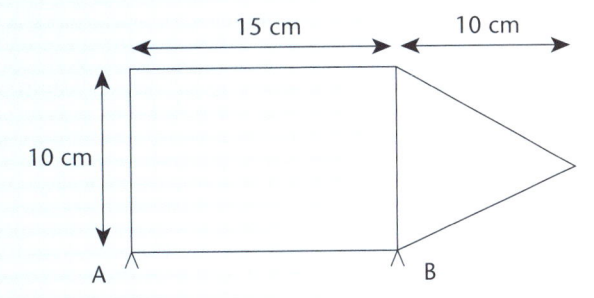

The diagram shows a uniform lamina of mass 0.8 kg resting on supports at A and B.

(a) Find the horizontal distance of the centre of mass from the support at A.

(b) Find the magnitude of the reaction force on the lamina at each support.

105

Mechanics 2

Practice examination questions (continued)

4

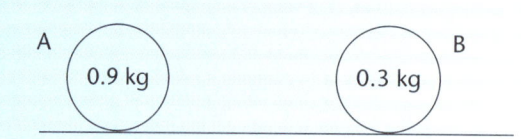

A (0.9 kg)   B (0.3 kg)

A and B are two smooth spheres of equal size and they are approaching each other on a direct line through their centres. A is moving with speed 4 ms^{-1} and B is moving with speed 2 ms^{-1}. The coefficient of restitution between the spheres is $\frac{1}{3}$.

Find:

(a) the speed of each sphere after the collision
(b) the impulse exerted by A on B during the collision
(c) the loss of mechanical energy during the collision.

5 Two particles P and Q approach each other with speeds u and $2u$ respectively. P has mass $3m$ and Q has mass m. The coefficient of restitution between the particles is e.

Given that the direction of motion of P is reversed in the collision, show that $e > \frac{1}{3}$.

6

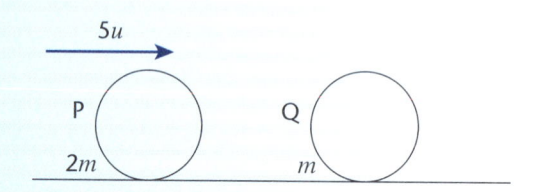

$5u \rightarrow$

P (2m)   Q (m)

A particle P of mass $2m$ moving with speed $5u$ strikes particle Q of mass m, which is at rest. After the collision both particles move in the same direction but the speed of particle Q is twice the speed of particle P.

(a) Find the speed of particle P after the collision.
(b) Find the magnitude of the impulse exerted on Q by P during the impact.

Mechanics 2

Practice examination questions (continued)

7. A motorbike is driven round a corner in a circular path of radius 40 m. The maximum speed that this can be done without skidding is 15 ms^{-1}.

   Calculate the coefficient of friction between the tyres and the road surface to 2 s.f.

   Take $g = 9.8$ ms^{-2}.

8. One end of a light inelastic string of length 50 cm is attached to a fixed point 40 cm above a smooth horizontal table. The other end is attached to an object of mass 2 kg moving in a circular path of radius 30 cm on the table.

   Given that the angular speed of the object is 3 radians per second, find:

   (a) the magnitude of the acceleration of the object
   (b) the tension in the string
   (c) the reaction force of the table on the object.

9. A particle P of mass m is attached to one end of a light inelastic string of length l. The other end of the string is attached to a fixed point O. P is released from rest with the string taut and horizontal.

   (a) Explain why, in the subsequent motion, the tension in the string does no work on the particle.
   (b) Explain why it is not valid to use the formula $v^2 = u^2 + 2as$ in this situation.
   (c) Find the maximum speed of the particle assuming that there is no resistance to motion.

10. A car of mass 900 kg travels up a straight road inclined at angle α to the horizontal, where $\sin \alpha = \frac{1}{15}$. Assume that the total non-gravitational force opposing the car's motion has a constant value of 800 N. Take $g = 9.8$ ms^{-2}.

    (a) Find the driving force required for the car to maintain a steady speed up the slope.
    (b) Find the power output of the car's engine given that its speed up the slope has a constant value of 20 ms^{-1}.
    (c) At the top of the hill, the road surface is horizontal. If the power output of the engine remains the same, what is the initial acceleration of the car?

11. A football of mass 0.2 kg is travelling horizontally towards the goal with speed 20 ms^{-1} when it is struck by the keeper. It then rebounds vertically with speed 5 ms^{-1}.

    Find the magnitude and direction of the impulse given to the ball by the keeper.

Practice examination questions (continued)

12 A car is driven round a track in a horizontal circle of radius 80 m at a speed of 20 ms^{-1}. The track is banked at angle α to the horizontal and there is no tendency for the car to side-slip.

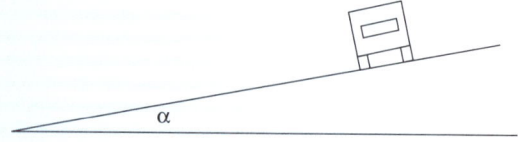

By representing the car as a particle:

(a) Find α to 1 d.p.

(b) The maximum speed that the car can travel on the banked track without slipping sideways is 28 ms^{-1}. Find the coefficient of friction between the tyres and the surface of the track.

13 The diagram shows a particle of mass 0.2 kg attached to one end of a light inelastic string of length 80 cm.

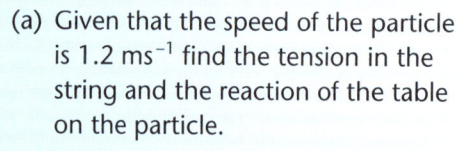

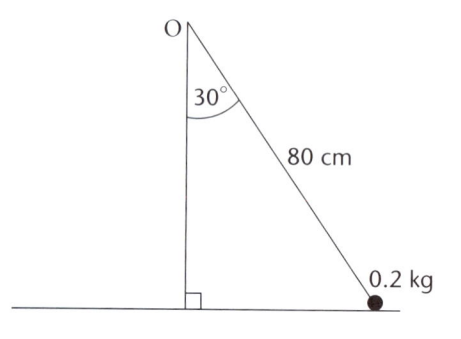

The particle moves in a horizontal circle on a smooth surface with the string inclined at 30° to the vertical.

(a) Given that the speed of the particle is 1.2 ms^{-1} find the tension in the string and the reaction of the table on the particle.

(b) Find the minimum angular speed of the particle so that the reaction of the table on the particle is zero.

Chapter 4
Decision Mathematics 2

The following topics are covered in this chapter:

- Game theory
- Networks
- Critical path analysis
- Matching and allocation
- Dynamic programming

4.1 Game theory

LEARNING SUMMARY

After studying this section you should be able to:

- understand the idea of a zero-sum game and its representation using a pay-off matrix
- identify play-safe strategies and stable solutions
- find optimal mixed strategies for a game with no stable solution
- reduce a pay-off matrix using a dominance argument

A two person zero-sum game

The starting point in the mathematical theory of games is that the outcome of a game is determined by the **strategies** of the players.

A two person **zero-sum game** is a game in which the winnings of one player equal the losses of the other for every combination of strategies. Taking winnings to be positive and losses to be negative gives a zero sum in each case.

Viewing a game from one player's point of view, we could represent the outcomes (called pay-offs) for each combination of strategies in a matrix. This is called the **pay-off matrix** for that player.

Example

A and B are two players in a zero-sum game. A uses one of two strategies, W or X, and B uses one of the strategies Y or Z. The table shows the pay-off matrix for A.

The situation could equally be represented by the pay-off matrix for B. This would show corresponding values with opposite signs since this is a zero-sum game.

Pay-off matrix for A		B	
		Y	Z
A	W	2	−2
	X	5	−4

The pay-off matrix shows that if B adopts strategy Y then the pay-off for A will be 2 by using strategy W and 5 using strategy X. On the other hand, if B adopts strategy Z then the pay-off for A will be −2 using strategy W and −4 using strategy X.

The idea is that neither player knows in advance which strategy the other will use.

Decision Mathematics 2

The play-safe strategy

The **play-safe** strategy for a player is the strategy for which the minimum pay-off is as high as possible. In the example above, the minimum pay-off for A using strategy W is −2, whereas the minimum pay-off using strategy X is −4. This means that strategy W is the play-safe strategy for player A.

Notice that finding the play-safe strategy for player A involves comparing the minimum values in the *rows* of the pay-off matrix for A. Finding the play-safe strategy for B will involve comparing the values in the *columns*, remembering that B's pay-offs are the negatives of the ones in the pay-off matrix for A.

The minimum value for B using strategy Y is −5 and the minimum value using strategy Z is 2. This means that the play-safe strategy for B is strategy Z.

The situation is shown in the pay-off matrices for A and B.

Pay-off matrix for A		B	
		Y	Z
A	W	2	−2
	X	5	−4

Pay-off matrix for B		B	
		Y	Z
A	W	−2	2
	X	−5	4

In this case, the maximum of the minimum pay-offs, for each player, is in the corresponding position in the two matrices. This represents the **stable solution** to the problem referred to as the **saddle point** (or **minimax point**).

The solution is stable in the sense that neither player can improve their pay-off by taking a different strategy, given that the other player doesn't change. In other words, while B uses strategy Z, the best strategy for A is W and while A uses strategy W, the best strategy for B is Z.

> In the example, the sum of these values is −2 + 2 = 0 and the solution is stable.

> **KEY POINT**
> If the sum of the two values used to determine the play-safe strategies is **not zero** then the values cannot correspond to the same cell position in the play-off matrices. This means that there is no saddle point and the game has **no stable solution**.

Example

The pay-off matrices for two players in a zero-sum game are given below. Show that there is no stable solution for the game.

Pay-off matrix for P		Q	
		Y	Z
P	W	5	−3
	X	−4	6

Pay-off matrix for Q		Q	
		Y	Z
P	W	−5	3
	X	4	−6

The play-safe strategies for P and Q are shown shaded.
The values used to determine the play-safe strategies are −3 and −5.
There is no stable solution since $(-3) + (-5) \neq 0$.

Decision Mathematics 2

Optimal strategies for games that are not stable

The repeated use of the same strategy over a series of games is called a **pure strategy**. This provides the best results for both players in a game which has a stable solution. In the case where no stable solution exists, a **mixed strategy** is used in which each of the strategies is employed with a given probability to find the optimal solution.

Returning to the previous example:

Pay-off matrix for P	Q	
	Y	Z
P W	5	−3
P X	−4	6

Suppose that P chooses:
    strategy W with probability p
and   strategy X with probability $(1 − p)$.

Then if:
Q chooses strategy Y, the expected *gain* for P is given by $5p − 4(1 − p) = 9p − 4$
Q chooses strategy Z, the expected *gain* for P is $−3p + 6(1 − p) = −9p + 6$

> The optimal value occurs when these expressions are equal.

$$9p − 4 = −9p + 6 \Rightarrow 18p = 10$$
$$\Rightarrow p = \tfrac{5}{9}$$

> You could equally use $−9p + 6$ to get the value of the game.

The value of the game is given by $9p − 4 = 1$.

Suppose that Q chooses:
    strategy Y with probability q
and   strategy Z with probability $(1 − q)$.

Then if:
P chooses strategy W, the expected *loss* for Q is given by $5q − 3(1 − q) = 8q − 3$
P chooses strategy X, the expected *loss* for Q is $−4q + 6(1 − q) = −10q + 6$

$$8q − 3 = −10q + 6 \Rightarrow 18q = 9$$
$$\Rightarrow q = \tfrac{1}{2}$$

(As a check, the value of the game is $8q − 3 = 1$ as before).

This shows that the optimal strategy for both players is to use mixed strategies such that:

P chooses strategy W with probability $\tfrac{5}{9}$ and strategy X with probability $\tfrac{4}{9}$

Q chooses strategy Y and strategy Z with equal likelihood.

The expected long-term gain then for P, as an average per game, is 1 and this is also the expected long-term loss, as an average per game, for Q.

Decision Mathematics 2

Graphical representation

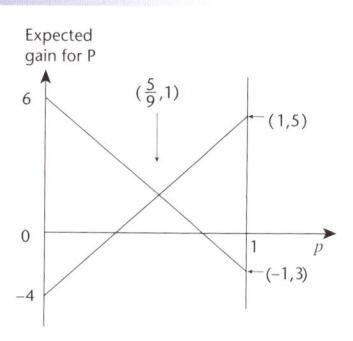

Each graph corresponds to a strategy for Q (i.e. the opponent of P).

The diagram shows graphs of $9p - 4$ and $-9p + 6$ against values of p from 0 to 1.

The point of intersection corresponds to the probability that gives the optimal mixed strategy for P in the last example.

> **KEY POINT**
>
> When P's opponent has more strategies there will be more graphs with several points of intersection. You will need to identify the one that represents the optimal mixed strategy for P.
> *This will be the highest point on or below each of the graphs.*

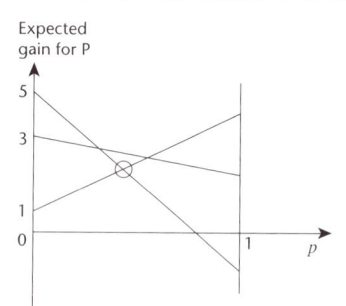

This diagram represents a situation where P's opponent has three strategies to choose from.

The point representing the optimal mixed strategy for P is circled.

Notice how the problem of identifying the right vertex can be expressed as a linear programming problem in which the object is to maximise the expected gain for P subject to the constraints represented by the regions bounded by the straight line graphs.

This is, in fact, the approach used for higher dimensional problems. The conditions are formulated as a linear programming problem which may then be solved by the Simplex algorithm.

> **KEY POINT**
>
> If the pay-offs for one strategy are *always* better than the corresponding pay-offs for some other strategy then the weaker one can be ignored when determining the probabilities for a mixed strategy. In this way, the pay-off matrix is reduced by what is called a **dominance argument**.

Progress check

1 The table shows a pay-off matrix for player A in a zero-sum game.

   (a) Find the play-safe strategy for each player.
   (b) Show that this game has no stable solution.
   (c) Find the best strategy for each player and the value of the game.

Pay-off matrix for A		Q	
		Y	Z
A	W	6	−4
	X	−3	5

1 (a) A(strategy X), Q(strategy Z)
   (b) $(-3) + (-5) \neq 0$
   (c) A use W and X with probabilities $\frac{4}{9}$ and $\frac{5}{9}$
       Q use Y and Z with equal probability, value of game = 1

Decision Mathematics 2

4.2 Networks

After studying this section you should be able to:

- understand flows in networks and be able to find the maximum flow for a network involving multiple sources and sinks

Flows in networks

The **flows** referred to may be flows of liquids, gases or any other measurable quantities. The edges may represent such things as pipes, wires or roads that carry the quantities between the points identified as vertices.

A typical vertex has a flow into it and a flow out of it. The exceptions are a **source** vertex which has no input and a **sink** vertex which has no output.

Each edge of the network has a **capacity** which represents the maximum possible flow along that edge. In the usual notation, the capacity is written next to the edge and the flow is shown in a circle.

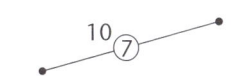

This edge has a capacity of 10 and a flow of 7.

The set of flows for a network is **feasible** if:

- The total output from all source vertices is equal to the total input for all sink vertices.
- The input for each vertex other than a source or sink vertex is equal to its output.
- The flow along each edge is less than or equal to its capacity.

A **cut** divides the vertices into two sets, one set containing the source and the other containing the sink.

The **capacity of a cut** is equal to the sum of the capacities of the edges that cross the cut, *taken in the direction from the source set to the sink set*.

In the diagram, the capacity of cut (i) is $15 + 10 = 25$.

The capacity of cut (ii) is $8 + 14 = 22$.

Notice that, for the second cut, the capacity of 4 is not included because it does not go from the source set to the sink set.

Be careful to add the capacities and not the flows.

The maximum flow–minimum cut theorem

The **maximum flow–minimum cut theorem** states that the maximum value of the flow through a network is equal to the capacity of the minimum cut. (This is a little bit like saying that a chain is only as strong as its weakest link.)

It follows from the theorem that if you find a flow that is equal to the capacity of some cut then it must be the maximum flow that can be established through the network.

For example, the capacity of cut (ii) for this network has already been found to be 22. The circled figures show one way of achieving this flow and the theorem now tells us that this must be the maximum possible flow through the network.

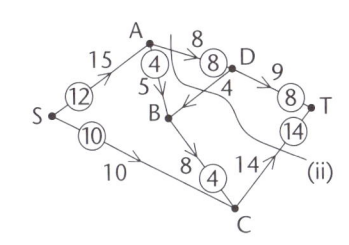

Decision Mathematics 2

Flow augmentation

Once an initial flow has been established through a network it is useful to know the extent to which the flow along any edge may be altered in either direction.

This is labelled as **excess capacity** and **back capacity** on each edge as shown. It may be possible to **augment** the flow through the network by making use of this flexibility along a path from the source vertex to the sink vertex. Such a path is called a **flow-augmenting path**.

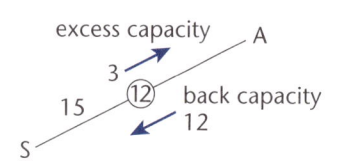

A **saturated** edge is one where the flow is equal to the capacity.

An algorithm for finding the maximum flow through a network by augmentation is:

Step 1 Find an initial flow by inspection.
Step 2 Label the excess capacity and back capacity for each edge.
Step 3 Search for a flow-augmenting path. If one can be found then increase the flow along the path by the maximum amount that remains feasible.
Step 4 Repeat steps 2 and 3 until no flow-augmenting path may be found.

Multiple sources and sinks

A network may have several sources S_1, S_2, ... and several sinks T_1, T_2, ... but the methods described in this section can still be applied through the use of a **supersource** and a **supersink**. A supersource S is an extra node, added to the network, that acts as a source for each of the original sources. Similarly, a supersink T is a node, added to the network, that acts as a sink for each of the original sinks.

Example

The network below shows two sources S_1 and S_2 and two sinks T_1 and T_2. Find the maximum flow through the network and state any flow-augmenting paths.

Explain why the flow is maximal.

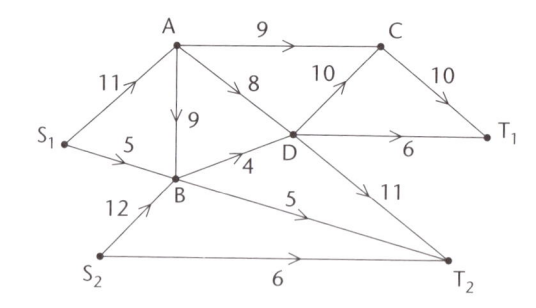

The first step is to add the supersource S and the supersink T to the network.

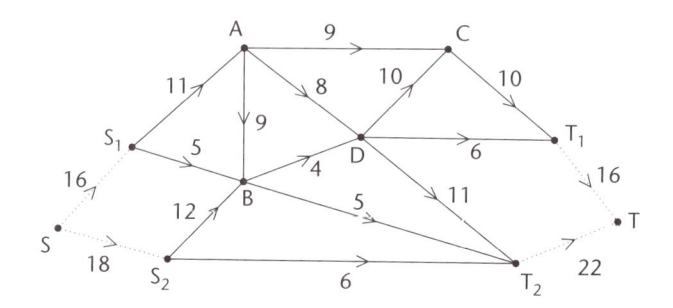

Decision Mathematics 2

The usual convention is to use dotted lines for the edges linking S and T to the network.

The capacities on the edges SS_1 and SS_2 have been chosen to support the maximum possible flows from S_1 and S_2. Similarly, the capacities on the edges T_1T and T_2T have been chosen to support the maximum flows into T_1 and T_2.
The diagram also shows a cut with capacity $11 + 4 + 5 + 6 = 26$.

A flow of 26 can be achieved as shown in the table.

Flow-augmenting path	Flow	Saturated edges
SS_1ACT_1T	9	S_1A, AC
SS_1ADT_1T	2	S_1A, DT_1
SS_2BDT_1T	4	BD, DT_1
SS_2BT_2T	5	BT_2
SS_2T_2T	6	S_2T_2
Total	26	

The flows on the flow-augmenting paths are shown circled on the network below. The flow of 26 is maximal since it equals the capacity of the cut (maximum flow minimum cut theorem).

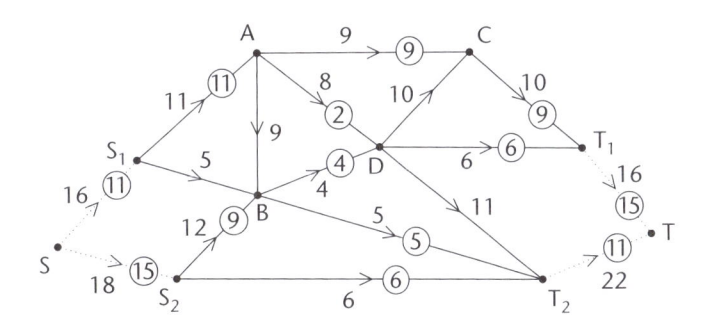

Note that each of the edges on the minimum cut is saturated. This will always be the case.

The final stage is to remove the supersource and supersink, along with the extra arcs, to show the maximum flow pattern for the original network.

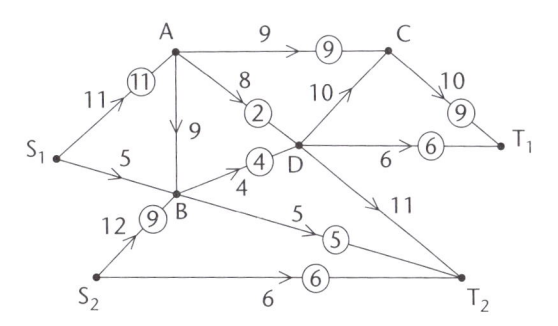

Decision Mathematics 2

Progress check

1. Find the maximum flow through this network. Verify your answer using the maximum flow–minimum cut theorem.

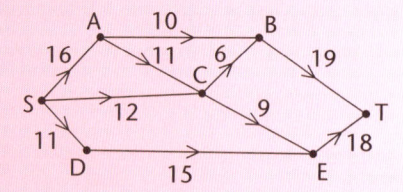

2. Use a supersource and supersink to find the maximum flow through this network.
   Establish the flow through B in this case.

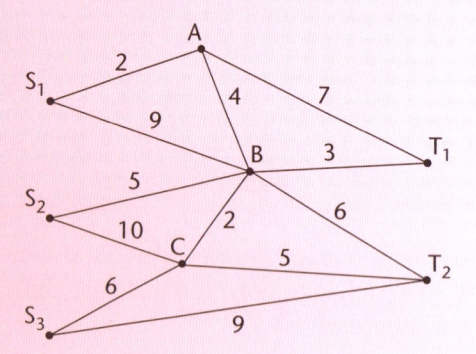

1. 34
2. 29, 13

Decision Mathematics 2

4.3 Critical path analysis

After studying this section you should be able to:

- construct an activity network from a given precedence table including the use of dummies where necessary
- use forward and backward scans to determine earliest and latest event times
- find the critical path in an activity network
- calculate the total float of an activity
- construct a Gantt chart for the purpose of scheduling
- draw a resource histogram and use resource levelling

The process of representing a complex project by a network and using it to identify the most efficient way to manage its completion is called **critical path analysis**.

Activity networks

A complex project may be divided into a number of smaller parts called **activities**. The completion of one or more activities is called an **event**.

Activities often rely on the completion of others before they can be started.

The relationship between these activities can be represented in a **precedence table**, sometimes called a **dependency table**.

In the precedence table shown on the right, the figures in brackets represent the **duration** of each activity, i.e. the time required, in hours, for its completion.

A precedence table can be used to produce an **activity network**. In the network, activities are represented by arcs and events are represented by vertices.

The vertices are numbered from 0 at the **start vertex** and finishing at the **terminal vertex**.

The direction of the arrows shows the order in which the activities must be completed.

There must only be *one* activity between each pair of events in the network. The notation (i, j) is used to represent the activity between events i and j.

A **dummy activity** is one that has zero duration. A dummy is needed in this network to show that G depends on C whereas F depends on C and E.

Activity	Depends on
A(3)	–
B(5)	–
C(2)	A
D(3)	A
E(3)	B, D
F(5)	C, E
G(1)	C
H(2)	F, G

A dummy is shown with a dotted line. Its direction is important in defining dependency. In this case it shows that F depends on C not that G depends on E.

Earliest event times

The **earliest event time** for vertex i is denoted by e_i and represents the earliest time of arrival at event i with all dependent activities completed. These times are calculated using a **forward scan** from the start vertex to the terminal vertex.

Latest event times

The **latest event time** for vertex i is denoted by l_i and represents the latest time that event i may be left without extending the time for the project. These times are

117

Decision Mathematics 2

calculated using a **backward scan** from the terminal vertex back to the start vertex. The **critical path** is the longest path through the network. The activities on this path are the **critical activities**. If any critical activity is delayed then this will increase the time needed to complete the project. The events on the critical path are the **critical events** and for each of these $e_i = l_i$.

It is useful to add the information about earliest and latest times to the network. The critical path is then easily identified.

Notice that B(5) lies between two critical events but it is not a critical activity.

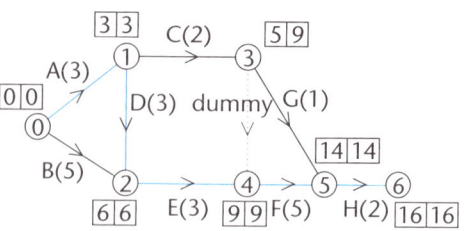

> The total float of any critical activity is always zero.

The **total float of an activity** is the maximum time that the activity may be delayed without affecting the length of the critical path. It is given by:

latest finish time − earliest start time − duration of the activity.

Scheduling

The process of allocating activities to workers for completion, within all of the constraints of the project, is known as **scheduling**.

The information regarding earliest and latest times for each activity is crucial when constructing a schedule. This information may be presented as a table or as a chart.

> Typically, the purpose of scheduling is to determine the number of workers needed to complete the project in a given time, or to determine the minimum time required for a given number of workers to complete the project.

Activity	Duration	Start Earliest	Start Latest	Finish Earliest	Finish Latest	Float
A(0, 1)	3	0	0	3	3	0
B(0, 2)	5	0	1	5	6	1
C(1, 3)	2	3	7	5	9	4
D(1, 2)	3	3	3	6	6	0
E(2, 4)	3	6	6	9	9	0
F(4, 5)	5	9	9	14	14	0
G(3, 5)	1	5	13	6	14	8
H(5, 6)	2	14	14	16	16	0

The critical activities are shown along one line.

> The diagram shown here is known as a cascade chart or Gantt chart.

The diagram illustrates the degree of flexibility in starting activities B, C and G. Remember that G cannot be started until C has been completed.

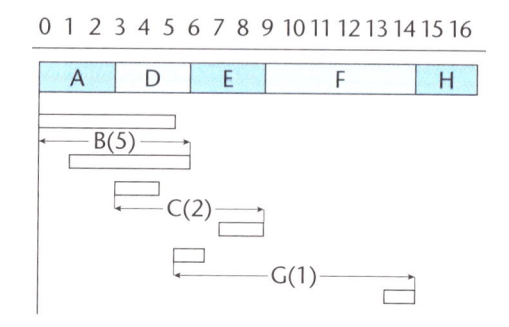

118

Decision Mathematics 2

Resource histograms and resource levelling

A **resource histogram** shows the demand on resources over the time taken to complete a project.

The example below shows how the number of workers needed varies from day to day over a period of 14 days.

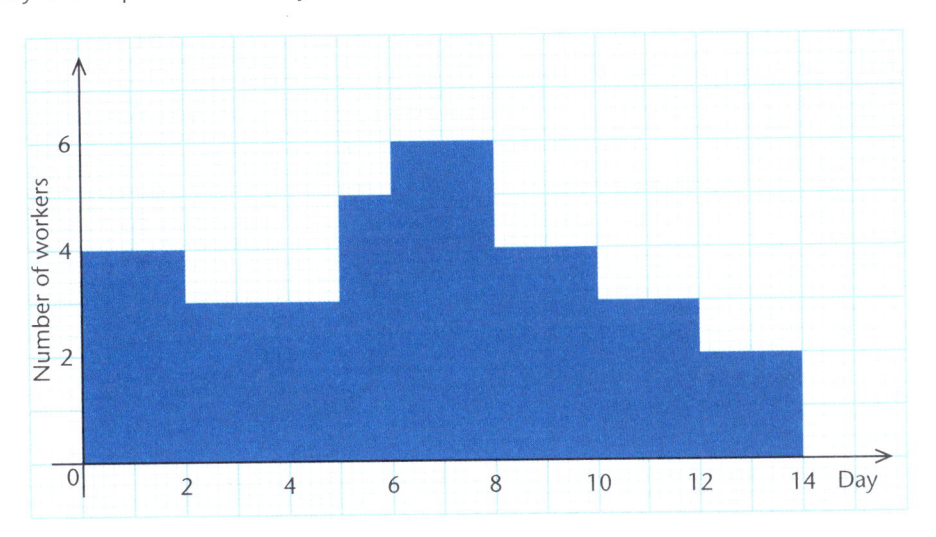

The maximum number of workers needed is 6 for days 7 and 8.

The diagram below shows the same resource histogram but includes details of the duration for each activity and the number of workers. Activity F, for example, lasts for 3 days and requires 2 workers. The critical activities are A, C, D, E and G. These are put in first and the rest of the diagram is built in layers on top.

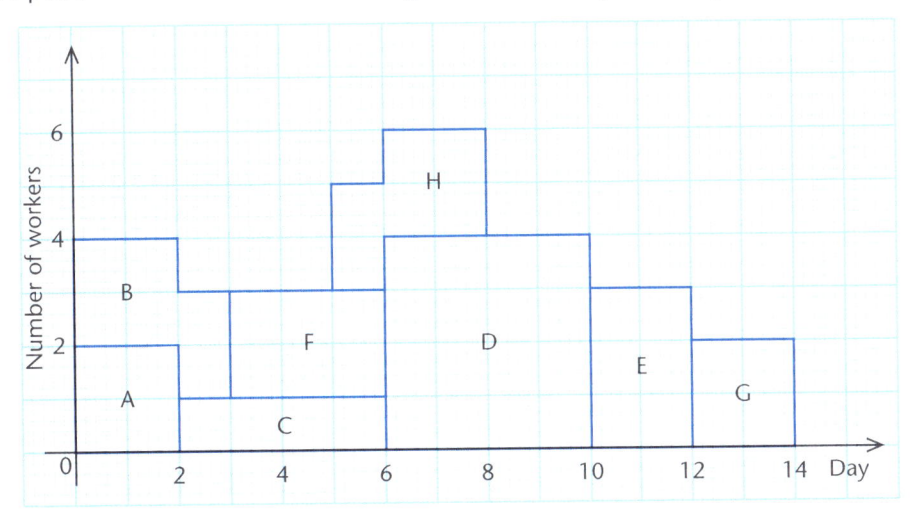

Once the resource histogram has been drawn, you may be asked to use **resource levelling** to work out the effect on the schedule of having a reduction in resources. For example, if the maximum number of workers available at any time is 5 then we can see from the diagram that this presents a problem for the completion of activity H.

Resource levelling involves selecting one or more activities to be delayed by an amount of time that allows the demand on resources to be brought down to the required level. The consequences of each delay must then be taken into account to determine the amount of extra time required.

In this example, delaying the start of activity D by 2 days allows activity H to be completed. Activities E and G must then also be delayed by 2 days and so the whole project requires 2 more days for completion.

Decision Mathematics 2

Progress check

1. Determine the critical activities and the length of the critical path for this network.

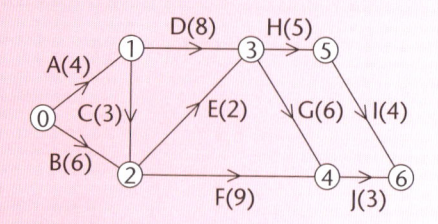

   [Network diagram with activities: A(4), B(6), C(3), D(8), E(2), F(9), G(6), H(5), I(4), J(3) between nodes 0,1,2,3,4,5,6]

2. Here is the activity network used in the example on page 118.

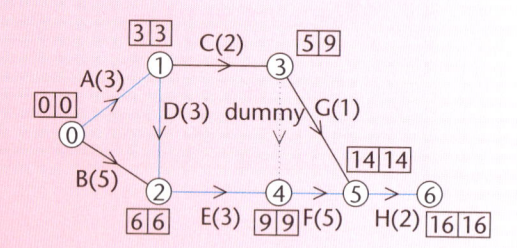

   [Network diagram with activities A(3), B(5), C(2), D(3), dummy, E(3), F(5), G(1), H(2); earliest/latest times: 0|0, 3|3, 5|9, 6|6, 9|9, 14|14, 16|16]

   The number of workers required for each activity is shown in the table.

A	B	C	D	E	F	G	H
2	3	4	1	2	3	2	2

   Assume that all workers are equally skilled and that once an activity has been started it must be completed without a break. Assume also that each activity starts at its earliest time.

   (a) Draw the resource histogram for the project assuming that the number of workers available is unlimited.

   (b) Given that the maximum number of workers available at any time is 5, use resource levelling to find the extra time needed to complete the project.

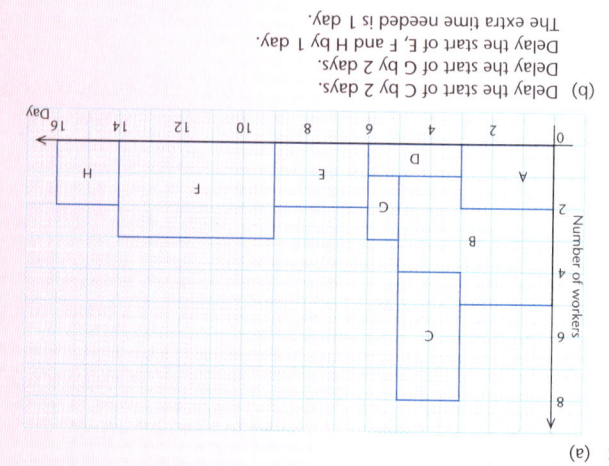

Answers:

1. A, D, G and J, 21

2. (a) [Resource histogram shown]

   (b) Delay the start of C by 2 days.
   Delay the start of G by 2 days.
   Delay the start of E, F and H by 1 day.
   The extra time needed is 1 day.

Decision Mathematics 2

4.4 Matching and allocation

After studying this section you should be able to:

- use bipartite graphs to model matchings
- understand the conditions for matchings to be maximal or complete
- apply the maximum matching algorithm

Matchings and graphs

A **bipartite graph** is a graph in which the vertices are divided into two sets such that no pair of vertices in the same set is connected by an edge.

In this case, the two sets are {A, B, C} and {p, q, r, s}.

Some vertices in a bipartite graph may not be connected to another vertex.

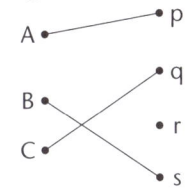

A **matching** between two sets may be represented by a bipartite graph in which there is at most one edge connecting a pair of vertices.

A **maximal matching** is a matching which has the maximum number of edges. This occurs when every vertex in one of the sets is connected to a vertex in the other set. The bipartite graph shown above represents a maximal matching.

A **complete matching** is a matching in which every vertex is connected to another vertex. This can only occur when the two sets contain the same number of vertices.

The matching improvement algorithm

Figure 1 is a bipartite graph showing the possible connections between two sets. It does not represent a matching because some vertices have more than one connection.

Figure 2 is a bipartite graph representing an initial matching.

An initial matching may be improved by increasing the number of connections. This is the purpose of the **matching improvement algorithm**.

Figure 1   Figure 2

> In an alternating path, the edges alternate between those from the bipartite graph that are not in the initial matching and those that are.

Step 1 In the initial matching, start from a vertex that is not connected and look for an **alternating path** to a vertex in the other set that is not connected.

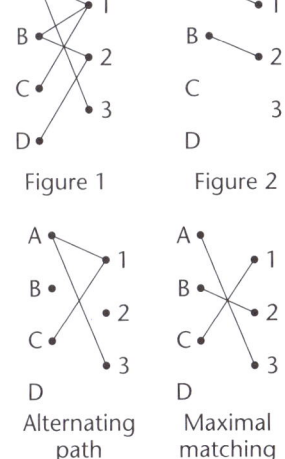

Alternating path   Maximal matching

Step 2 Each edge on the alternating path, not included in the initial matching, is now included and each edge originally included is removed.

Step 3 Repeat steps 1 and 2 using the latest matching in place of the initial matching until no further alternating paths can be found.

Allocation problems

In an allocation (or assignment) problem, the object is to set up a matching that will optimise a particular objective, such as minimising a cost or maximising a

Decision Mathematics 2

profit. If we regard one set of objects as agents and the other set as tasks, then the underlying assumptions are that:

- Any agent may be assigned to any task (unlike the matching problems).
- Each agent is assigned at most one task and each task has at most one agent assigned to it.

When the number of agents is equal to the number of tasks, the problem is said to be **balanced**. If the number of agents exceeds the number of tasks, or vice versa, then the problem is **unbalanced**. The technique for solving an unbalanced problem involves introducing a **dummy** agent or task which then allows the solution to proceed as for the balanced case. In the solution of an unbalanced problem, a task assigned to a dummy agent isn't actually completed and an agent assigned to a dummy task doesn't actually have a task to carry out.

The opportunity cost matrix

The table shows the time in hours for three workers A, B and C to complete three separate tasks.

		Task		
		1	2	3
Worker	A	9	11	7
	B	8	10	9
	C	7	9	8

The **cost matrix** shows only the figures in hours. In this case, the 'cost' is measured in time.

9	11	7
8	10	9
7	9	8

The **opportunity cost matrix** is found from the cost matrix by:

1. Subtracting the smallest value in each row from every other value in that row.
2. Subtracting the smallest value in each column from every other value in that column.

9	11	7
8	10	9
7	9	8

→ 1

2	4	0
0	2	1
0	2	1

↓ 2

2	2	0
0	0	1
0	0	1

The Hungarian algorithm

The **Hungarian algorithm** is used to find the allocation that optimises the given objective.

Step 1 Find the opportunity cost matrix.

Step 2 Test for optimality. If true then make the allocation and stop.

Step 3 Revise the opportunity cost matrix and repeat step 2.

Decision Mathematics 2

The method used to test for optimality is to find the minimum number of horizontal and/or vertical lines needed to cover all of the zeros in the opportunity cost matrix. If this number = n where the matrix has n rows and n columns, then the matrix will give an optimal allocation.

In the example, the number of lines needed = 3.
The matrix has 3 rows and columns so it will give an optimal allocation.

2	2	0
0	0	1
0	0	1

The optimal allocation is now found by selecting positions of zeros in the opportunity cost matrix such that one zero is selected in each row and in each column.

One possibility is shown here along with the corresponding figures in the original cost matrix.

2	2	**0**
0	**0**	1
0	0	1

9	11	**7**
8	**10**	9
7	9	8

The allocation is A → 3, B → 2 and C → 1 giving a total cost, in hours, of $7 + 10 + 7 = 24$.

In this case, there is an alternative solution.

2	2	**0**
0	0	1
0	**0**	1

9	11	**7**
8	10	9
7	**9**	8

The allocation is A → 3, B → 1 and C → 2 giving a total cost, in hours, of $7 + 8 + 9 = 24$.

Revising the opportunity cost matrix

When the opportunity cost matrix does not give an optimal solution, a revised opportunity cost matrix is found as follows:

Step 1 Find the smallest number not covered by a line. Subtract this number from every number not covered by a line.

Step 2 Add this number to every number lying at the intersection of two lines.

For example, this opportunity cost matrix has 3 rows and columns, but its zeros can be covered with just 2 lines.

0	3	1
0	2	5
0	0	0

Applying the steps given above produces this revised opportunity cost matrix.

0	2	**0**
0	1	4
1	**0**	0

123

Decision Mathematics 2

Maximising allocation problems

If the objective involves maximising a quantity, such as a profit, then the first step is to subtract every value in the original matrix from a fixed value. This fixed value may be taken to be the largest value in the matrix or any value larger than this. The Hungarian algorithm is now applied in the usual way to this transformed matrix.

Progress check

1. Starting with an initial matching in which 1 is connected to R and 2 is connected to P, use the improvement algorithm to establish a complete matching for this bipartite graph.

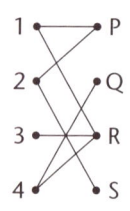

2. Given the cost matrix:

	1	2	3
A	3	3	4
B	8	5	9
C	2	1	4

Use the Hungarian algorithm to find the minimum total cost.

1 1–P, 2–S, 3–R, 4–Q
2 4 + 5 + 2 = 11

Decision Mathematics 2

4.5 Dynamic programming

LEARNING SUMMARY

After studying this section you should be able to:

- understand the concept of dynamic programming
- set up a dynamic programming tabulation and use it to solve a problem

Dynamic programming

Dynamic programming is one of the techniques used to solve problems where some quantity is to be optimised. It involves identifying **stages** and all possible **states** and **actions** within a process. The stages are counted from the end of the process to the beginning.

Working backwards through the stages, decisions are made at each stage that optimise the quantity. Importantly, the calculations carried out must follow from the corresponding optimal value found at the previous stage. In effect, this is applying the **principle of optimality** which states that **any part of an optimal path is itself optimal**.

In the examination, information about the process and the quantity to be optimised may be given in the form of a network or a table.

Example

The network below shows the possible actions and corresponding anticipated profits for a company over the next five months. The figures represent the profit in units of ten thousand pounds, with a negative value indicating a loss due to investment.

(a) Use dynamic programming to maximise the expected profit over the five months.
(b) Give the sequence of vertices in order to achieve the maximum profit.

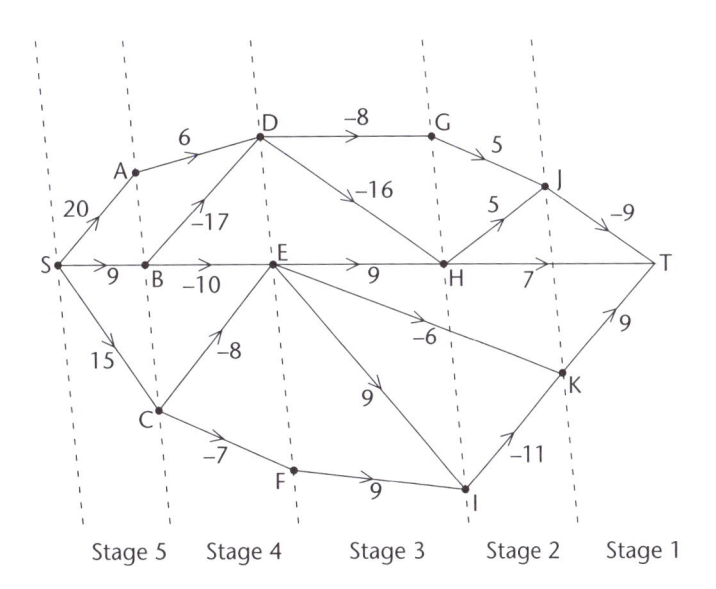

Again, in the examination, an insert is usually provided showing a table to be completed. The table on the next page is typical of the ones used and the coloured entries are typical of the information that would be given on the insert.

(a) An asterisk in the right-hand column is used to indicate the optimal value from the given state.

Stage	State	Action	Value	
1	J	JT	–9	*
	K	KT	9	*
2	G	GJ	5 + –9 = –4	*
	H	HJ	5 + –9 = –4	
		HT	7	*
	I	IK	–11 + 9 = –2	*
3	D	DG	–8 + –4 = –12	
		DH	–16 + 7 = –9	*
	E	EH	9 + 7 = 16	*
		EK	–6 + 9 = 3	
		EI	9 + –2 = 7	
	F	FI	9 + –2 = 7	*
4	A	AD	6 + –9 = –3	*
	B	BD	–17 + –9 = –26	
		BE	–10 + 16 = 6	*
	C	CE	–8 + 16 = 8	*
		CF	–7 + 7 = 0	
5	S	SA	20 + –3 = 17	
		SB	9 + 6 = 15	
		SC	15 + 8 = 23	*

JT and KT are the only possible paths from J and K, so they are automatically optimal.

The value for GJ is calculated as 5 (shown on GJ) plus –9 (the optimal value from J). This is the only value from G so it is automatically optimal.

There are two possible moves from H. The optimal value is 7.

There are three possible moves from E. The value for EH, for example, is 9 (shown on EH) plus 7 (the optimum value for H). This gives 16, which turns out to be the optimal value from E.

F is the last vertex to consider from Stage 3.

The three vertices to consider in Stage 4 are A, B and C.

The highest value found at the final stage is the overall maximum value. In this case this is 23 which represents 23 × £10 000.

The maximum expected profit over the next five months is £230 000.

(b) The sequence of vertices that maximises the profit is SCEHT.

Different types of problem

The dynamic programming technique may be used to solve various types of problem. The way in which the table is used remains the same, but the way in which the optimum value is selected may be different. To find the minimum distance through a network, for example, select the minimum value for each state rather than the maximum as in the example above.

In particular, two types that you need to be aware of are the **maximin** and **minimax**.

The optimal values for the maximin are found by first selecting the minimum value for each action and then choosing the maximum of these to represent the state.

Decision Mathematics 2

To illustrate how the maximin works, here is part of a network and the corresponding part of a table.

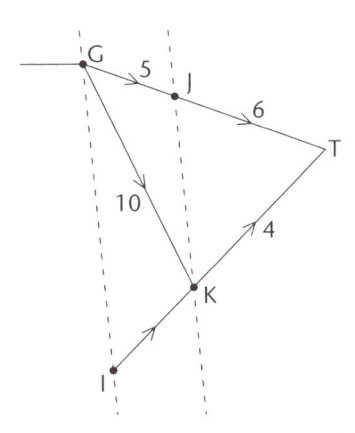

Stage	State	Action	Value	
1	J	JT	6	*
	K	KT	4	*
2	G	GJ	Min(5, 6) = 5	*
		GK	Min(10, 4) = 4	

JT and KT are the only possible paths from J and K, so they are automatically optimal.

The value for GJ is the minimum of 5 (shown on the arc GJ) and 6 (the optimum from J at the previous stage).

The value for GK is the minimum of 10 (shown on the arc GJ) and 4 (the optimum from K at the previous stage).

There are two values for the state G and the maximum of these (5) is shown with an asterisk as this is the optimum for G.

In a similar way, the optimal values for the minimax are found by first selecting the maximum value for each action and then choosing the minimum of these to represent the state.

The table shows the minimax values for the part of the network given above.

Stage	State	Action	Value	
1	J	JT	6	*
	K	KT	4	*
2	G	GJ	Max(5, 6) = 6	*
		GK	Max(10, 4) = 10	

Progress check

1. The diagram shows part of a network. Find the maximin value for the initial state A.

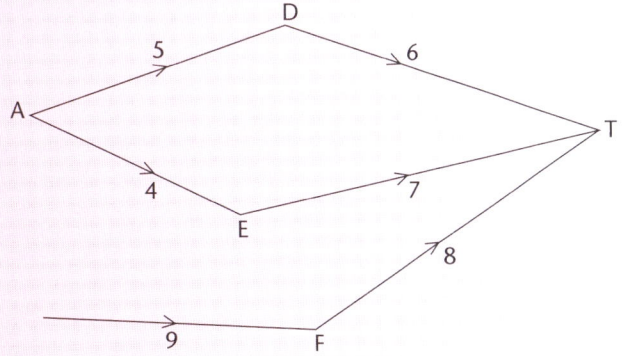

2. Find the minimax value for the initial state A in the diagram in question 1.

1 5
2 6

Decision Mathematics 2

Sample questions and model answers

1

Ashley and Emma play a two-person zero-sum game in which they each consider three possible strategies. The table shows the pay-off matrix for Ashley.

		Emma		
		X	Y	Z
Ashley	A	5	−3	3
	B	4	1	−2
	C	−3	6	−4

(a) Find the play-safe strategy for each player and show that the game does not have a stable solution.

(b) Set up a linear programming problem to find Ashley's optimal mixed strategy. You are not required to solve it.

(a) The minimum row values are −3, −2 and −4 in Ashley's pay-off matrix. The maximum of these is −2 and so the play-safe strategy for Ashley is strategy B.

The minimum column values in Emma's pay-off matrix are −5, −6 and −3. The maximum of these is −3 and so the play-safe strategy for Emma is strategy Z.

		Emma		
		X	Y	Z
	A	5	−3	3
Ashley	B	4	1	−2
	C	−3	6	−4

		Emma		
		X	Y	Z
	A	−5	3	−3
Ashley	B	−4	−1	2
	C	3	−6	4

The values used to determine the play-safe strategies do not have the corresponding positions in the pay-off matrices so there is no stable solution. Alternatively, $-3 + -2 \neq 0$ and so, again, there is no stable solution.

Decision Mathematics 2

Sample questions and model answers (continued)

The values need to be made positive by adding a fixed value to each one.

(b) Adding 5 to each of the figures in Ashley's pay-off matrix gives:

$$\begin{array}{ccc} 10 & 2 & 8 \\ 9 & 6 & 3 \\ 2 & 11 & 1 \end{array}$$

Using a, b and c to represent the probability that Ashley uses strategy A, B or C respectively, the linear programming problem is formulated as:

Maximise P subject to $a + b + c \leq 1$ where $a \geq 0$, $b \geq 0$, $c \geq 0$ and

$$P - 10a - 9b - 2c \leq 0$$
$$P - 2a - 6b - 11c \leq 0$$
$$P - 8a - 3b - c \leq 0$$

and P represents the value of the game.

2

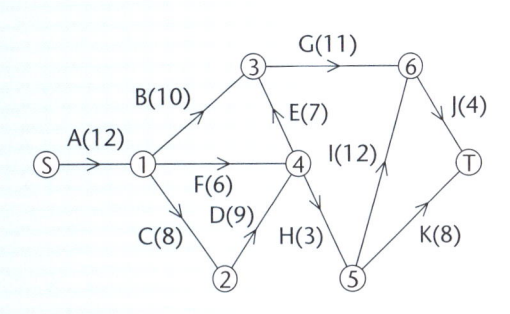

The diagram is an activity network for a project.

The figures in brackets represent the time in days to complete each activity.

The start vertex is S and the terminal vertex is T.

(a) Label each vertex with the earliest and latest time for the event. Give the length of the critical path.

(b) State which events are critical and give the critical path.

Use a forward scan to find the earliest event times.

Use a backward scan to find the latest event times.

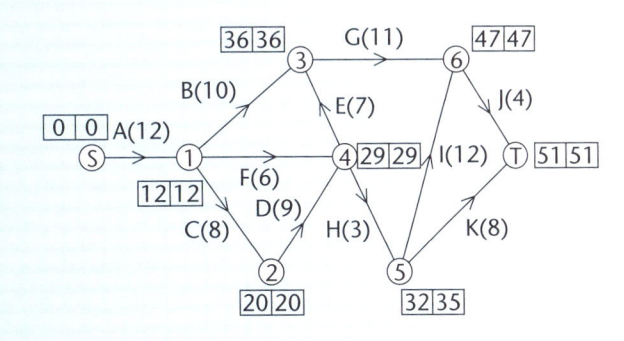

(a) The length of the critical path is 51 days.

(b) The critical events are S, 1, 2, 3, 4, 6 and T.
The critical path is A, C, D, E, G and J.

The length of the critical path is given by the earliest (and latest) event time at the terminal vertex.

Decision Mathematics 2

Practice examination questions

1 A and B are the players in a two-person zero-sum game. Each may use one of two strategies when a game is played. The pay-off matrix for A is shown below.

		Player B	
Player A		Y	Z
	W	4	8
	X	6	3

(a) Find the best mixed strategy for each player.

(b) Find the value of the game.

2 Richard and Judy play a zero-sum game. The pay-off matrix for Richard is shown in the table.

		Judy		
		X	Y	Z
Richard	A	4	3	−4
	B	−3	6	8
	C	2	1	−6

(a) Find the play-safe strategy for each player.

(b) Show that there is no saddle point for the game.

(c) Explain why strategy C will not be part of Richard's mixed strategy.

(d) Find Richard's optimal mixed strategy.

Decision Mathematics 2

Practice examination questions (continued)

3 The table shows the names of five employees of a company and the days when they are each available to work a late shift. The days that have been highlighted represent a first attempt at matching the employees to the available days to make a roster.

Name	Days available
Abbie	**Mon**, Tue, Fri
Ben	Mon, **Wed**, Thu
Colin	Mon, **Fri**
Dave	Fri, **Wed**
Emma	**Thu**, Fri

(a) Use a bipartite graph to represent the relationship between employees and days available.

(b) Use a second bipartite graph to represent an initial matching based on the days highlighted.

(c) Implement a matching improvement algorithm to obtain a complete matching. Indicate the alternating path used.

4 An expedition is planned across a desert from S to T. The network shows the possible routes and distances in km between places where water and supplies are available. The optimal route is the minimax route that minimises the maximum distance between supply points.

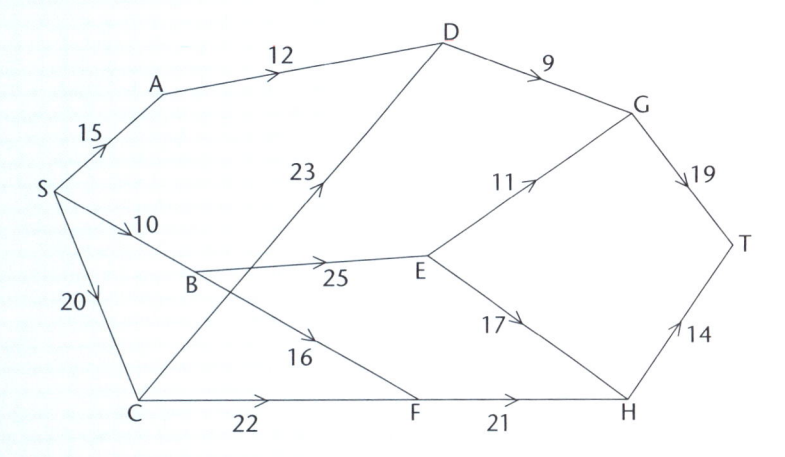

By completing the following table, or otherwise, use dynamic programming, working backwards from T, to find the minimax route and give the maximum distance between supply points on the route.

Stage	State	Action	Value	
1	G	GT	19	*
	H	HT	14	*
2	D	DG	Max(9, 19) = 19	
	E	EH		
	F	FH		
3	A	AD		
	B	BE		
		BF		
4				

131

Practice examination answers

Core 3 and Core 4 (Pure Mathematics)

1 $\dfrac{1}{(1+x)^2} = (1+x)^{-2}$

$= 1 + (-2)x + \dfrac{(-2)(-3)}{2!}x^2 + \ldots$

$= 1 - 2x + 3x^2 + \ldots \quad |x| < 1$

$\sqrt{4+x} = \sqrt{4\left(1+\dfrac{x}{4}\right)}$

$= 4^{\frac{1}{2}}\left(1+\dfrac{x}{4}\right)^{\frac{1}{2}}$

$= 2\left(1+\dfrac{x}{4}\right)^{\frac{1}{2}}$

$= 2\left(1 + (\tfrac{1}{2})\left(\dfrac{x}{4}\right) + \dfrac{(\tfrac{1}{2})(-\tfrac{1}{2})}{2!}\left(\dfrac{x}{4}\right)^2 + \ldots\right)$

$= 2 + \dfrac{x}{4} - \dfrac{x^2}{64} + \ldots \quad |x| < 4$

$\therefore \ f(x) = 1 - 2x + 3x^2 + 2 + \dfrac{x}{4} - \dfrac{x^2}{64} + \ldots$

$= 3 - 1\tfrac{3}{4}x + 2\tfrac{63}{64}x^2 + \ldots \quad |x| < 1$

Ignoring terms in x^3 and higher, $f(x) \approx a + bx + cx^2$, with $a = 3$, $b = -1\tfrac{3}{4}$, $c = 2\tfrac{63}{64}$.

The restriction is $|x| < 1$.

2 (a) $y = \cos^2 x \sin x$

$\dfrac{dy}{dx} = \cos^2 x(\cos x) + \sin x(-2\sin x \cos x)$

$= \cos^3 x - 2\cos x \sin^2 x$

$= \cos^3 x - 2\cos x(1 - \cos^2 x)$

$= \cos^3 x - 2\cos x + 2\cos^3 x$

$= 3\cos^3 x - 2\cos x$

(b) $u = \cos x \Rightarrow \dfrac{du}{dx} = -\sin x$

$\dfrac{dx}{du} = -\dfrac{1}{\sin x} \Rightarrow \sin x \dfrac{dx}{du} = -1$

$\int \cos^2 x \sin x \, dx$

$= \int \cos^2 x \sin x \dfrac{dx}{du} du$

$= \int (-u^2) du$

$= -\tfrac{1}{3} u^3 + c$

$= -\tfrac{1}{3} \cos^3 x + c$

(This can be also be done by direct recognition, see page 43.)

3 $\dfrac{dy}{dx} = \dfrac{2x(x+1) - 1(x^2 - 4)}{(x+1)^2}$

$= \dfrac{x^2 + 2x + 4}{(x+1)^2}$

When $x = 2$, $\dfrac{dy}{dx} = \dfrac{4}{3}$

$\therefore$ gradient of normal $= -\tfrac{3}{4}$

Equation of normal at $(2, 0)$:

$y - 0 = -\tfrac{3}{4}(x - 2)$

$4y = -3x + 6$

$3x + 4y - 6 = 0$

$\therefore \ a = 3$, $b = 4$ and $c = -6$.

4 (a) $g(x) = (x+1)(x-1) = x^2 - 1$.

Since $x^2 \geqslant 0$ the range of g is $x \geqslant -1$.

(b) $fg(x) = f(x^2 - 1)$
$= 5(x^2 - 1) - 7$
$= 5x^2 - 12$.

(c) $fg(x) = f(x)$
$\Rightarrow 5x^2 - 12 = 5x - 7$
$\Rightarrow 5x^2 - 5x - 5 = 0$
$\Rightarrow x^2 - x - 1 = 0$
$\Rightarrow x = \dfrac{1 \pm \sqrt{5}}{2}$.

5 (a) $R\sin(x + a) \equiv 15\sin x + 8\cos x$

$\Rightarrow R\cos a = 15$ and $R\sin a = 8$ (a is acute)

$\Rightarrow R = \sqrt{15^2 + 8^2} = 17$

$\tan a = \tfrac{8}{15} \Rightarrow a = 28.07°$

So $15\sin x + 8\cos x \equiv 17\sin(x + 28.07°)$.

(b) Max value is 17.

(c) $15\sin x + 8\cos x = 12$

$\Rightarrow 17\sin(x + 28.07°) = 12$.

$\sin^{-1}(\tfrac{12}{17}) = 44.90°$, $180° - 44.90° = 135.10°$

$x + 28.07 = 44.90° \Rightarrow x = 16.8°$

$x + 28.07 = 135.10° \Rightarrow x = 107.0°$

Practice examination answers

Core 3 and Core 4 (Pure Mathematics) (continued)

6 $x\dfrac{dy}{dx} = y + yx \Rightarrow x\dfrac{dy}{dx} = y(1+x)$

Separating the variables:
$\dfrac{1}{y}\dfrac{dy}{dx} = \dfrac{1+x}{x}$

$\dfrac{1}{y}\dfrac{dy}{dx} = \dfrac{1}{x} + 1$

$\int \dfrac{1}{y}\,dy = \int \left(\dfrac{1}{x} + 1\right)dx$

∴ $\ln y = \ln x + x + c$

When $x = 2, y = 4$,

$\Rightarrow \ln 4 = \ln 2 + 2 + c$

$\Rightarrow c = \ln 2 - 2 \quad (\ln 4 = 2\ln 2)$

So $\ln y = \ln x + x + \ln 2 - 2$

$\ln\left(\dfrac{y}{2x}\right) = x - 2$

$\dfrac{y}{2x} = e^{x-2}$

$y = 2xe^{x-2}$

7 (a) (i) $y = (4x-3)^9$

$\Rightarrow \dfrac{dy}{dx} = 9(4x-3)^8 \times 4$

$\Rightarrow \dfrac{dy}{dx} = 36(4x-3)^8.$

(ii) $y = \ln(5-2x)$

$\dfrac{dy}{dx} = \dfrac{1}{5-2x} \times (-2)$

$= \dfrac{-2}{5-2x}$ (you can write this as $\dfrac{2}{2x-5}$).

(b) (i) $\int (4x-3)\,dx = \dfrac{1}{36}(4x-3)^9 + C.$

(ii) $\int_3^4 \dfrac{1}{5-2x}\,dx = -\dfrac{1}{2}\int_3^4 \dfrac{-2}{5-2x}\,dx$

$= -\dfrac{1}{2}[\ln|5-2x|]_3^4$

$= -\dfrac{1}{2}(\ln|-3| - \ln|-1|)$

$= -\dfrac{1}{2}\ln 3.$

8 (a) $\int_0^1 xe^{-2x}\,dx = \left[x\left(-\tfrac{1}{2}e^{-2x}\right)\right]_0^1 - \int_0^1 \left(-\tfrac{1}{2}e^{-2x}\right)dx$

$= -\tfrac{1}{2}e^{-2} + \tfrac{1}{2}\int_0^1 e^{-2x}\,dx$

$= -\tfrac{1}{2}e^{-2} + \tfrac{1}{2}\left[\left(-\tfrac{1}{2}e^{-2x}\right)\right]_0^1$

$= -\tfrac{1}{2}e^{-2} + \tfrac{1}{2}(-\tfrac{1}{2}e^{-2} - (-\tfrac{1}{2}))$

$= -\tfrac{3}{4}e^{-2} + \tfrac{1}{4}$

(b) (i) $y = xe^{-x}$

$\dfrac{dy}{dx} = x(-e^{-x}) + e^{-x} = e^{-x}(1-x)$

$\dfrac{dy}{dx} = 0$ when $x = 1$

When $x = 1, y = e^{-1} \Rightarrow A$ is point $(1, e^{-1})$.

(ii) $V = \pi \int_0^1 y^2\,dx = \pi \int_0^1 x^2 e^{-2x}\,dx$

$= \pi\left(\left[x^2(-\tfrac{1}{2}e^{-2x})\right]_0^1 - \int_0^1 -\tfrac{1}{2}e^{-2x}(2x)\,dx\right)$

$= \pi\left(-\tfrac{1}{2}e^{-2} + \int_0^1 xe^{-2x}\,dx\right)$

$= \pi(-\tfrac{1}{2}e^{-2} - \tfrac{3}{4}e^{-2} + \tfrac{1}{4})$ (using part (a))

$= \dfrac{\pi}{4}(1 - 5e^{-2})$

9 $\dfrac{dP}{dt} \propto P \Rightarrow \dfrac{dP}{dt} = kP$, where $k > 0$.

Separating the variables

$\int \dfrac{1}{P}\,dP = \int k\,dt$

$\Rightarrow \ln P = kt + c$

$\Rightarrow P = e^{kt+c}$

$\Rightarrow P = Ae^{kt}$ (where $A = e^c$)

$t = 0, P = 500 \Rightarrow 500 = A$

$t = 10, P = 1000 \Rightarrow 1000 = 500e^{10k}$

∴ $2 = e^{10k}$

$\Rightarrow 10k = \ln 2$

$\Rightarrow k = 0.1 \ln 2$

∴ $P = 500e^{(0.1 \ln 2)t}$

When $t = 20$,

$P = 500e^{0.1 \ln 2 \times 20} = 500 \times 4 = 2000$

So the population is 2000 when $t = 20$.

133

Practice examination answers

Core 3 and Core 4 (Pure Mathematics) (continued)

10 (a) $x^3 + xy + y^2 = 7$

$3x^2 + x\dfrac{dy}{dx} + y + 2y\dfrac{dy}{dx} = 0$

At (1, 2)

$3 + \dfrac{dy}{dx} + 2 + 4\dfrac{dy}{dx} = 0$

$\Rightarrow 5\dfrac{dy}{dx} + 5 = 0$ i.e. $\dfrac{dy}{dx} = -1$

At (1, 2) gradient of normal = 1

Equation of normal is

$y - 2 = 1(x - 1)$

i.e. $y = x + 1$

(b) $x = t^2,\ y = t + \dfrac{1}{t}$.

$\dfrac{dx}{dt} = 2t$

$\dfrac{dy}{dt} = 1 - \dfrac{1}{t^2} = \dfrac{t^2 - 1}{t^2}$

$\dfrac{dy}{dx} = \dfrac{dy}{dt} \times \dfrac{dt}{dx} = \dfrac{t^2 - 1}{2t^3}$

$\dfrac{dy}{dx} = 0$ when $t^2 - 1 = 0 \Rightarrow t = \pm 1$

When $t = 1$, coordinates are (1, 2).
When $t = -1$, coordinates are (1, −2).
The stationary points are at (1, 2) and (1, −2).

11 (a) $\sin 2x = 0$ when $2x = 0,\ \pi,\ 2\pi, \ldots$
i.e. when $x = 0,\ \tfrac{1}{2}\pi,\ \pi, \ldots$
$\therefore a = \tfrac{1}{2}\pi$

(b) $A = \displaystyle\int_0^{\frac{1}{2}\pi} y\,dx = \int_0^{\frac{1}{2}\pi} \sin 2x\,dx$

$= -\tfrac{1}{2}\Big[\cos 2x\Big]_0^{\frac{1}{2}\pi}$

$= -\tfrac{1}{2}(-1 - 1)$

$= 1$

Area of R is 1 square unit.

(c) $V = \pi \displaystyle\int_0^{\frac{1}{2}\pi} y^2\,dx$

$= \pi \displaystyle\int_0^{\frac{1}{2}\pi} \sin^2 2x\,dx$

$= \tfrac{1}{2}\pi \displaystyle\int_0^{\frac{1}{2}\pi} (1 - \cos 4x)\,dx$

$= \tfrac{1}{2}\pi \Big[x - \tfrac{1}{4}\sin 4x\Big]_0^{\frac{1}{2}\pi}$

$= \tfrac{1}{2}\pi(\tfrac{1}{2}\pi - 0 - 0)$

$= \tfrac{1}{4}\pi^2$

The volume generated is $\tfrac{1}{4}\pi^2$ cubic units.

12 (a) Referring to origin (0, 0)
a = 2**i** + 2**j** and **b** = 2**i** + 3**j** + **k**

Also **b** − **a** = **j** + **k**

A vector equation for AB is

r = **a** + μ(**b** − **a**) where μ is a scalar parameter

$\therefore$ **r** = 2**i** + 2**j** + μ(**j** + **k**)

(b) If lines intersect, there are unique values for μ and λ for which

2**i** + 2**j** + μ(**j** + **k**) = 2**j** + **k** + λ(**i** + **k**)

Equating coefficients of **i**, **j** and **k**, in turn gives:

$2 = \lambda$

$2 + \mu = 2$

$\mu = 1 + \lambda$

There are no values of μ and λ that satisfy all three equations, so the lines have no common point.

(c) The angle between the lines is the angle between their direction vectors.

$|\mathbf{j} + \mathbf{k}| = \sqrt{2}$ and $|\mathbf{i} + \mathbf{k}| = \sqrt{2}$

$(\mathbf{j} + \mathbf{k}) \cdot (\mathbf{i} + \mathbf{k}) = 1$

Let the angle be θ, then

$\cos\theta = \dfrac{1}{\sqrt{2}\sqrt{2}} = 0.5$

$\Rightarrow \theta = 60°$

Practice examination answers

Probability and Statistics 2

1 If $\mu = 20$, $\bar{X} \sim N(20, \frac{16}{10})$,

i.e. $\bar{X} \sim N(20, 1.6)$.

Carrying out a two-tailed test at 5% level, reject H_0 if $z < -1.96$ where

$$z = \frac{17.2 - 20}{\sqrt{1.6}} = -2.213\ldots$$

Since $z < -1.96$, reject H_0.

There is evidence that μ is not 20.

2 (a) If X is number of people with the health problem, then $X \sim B(80, 0.01)$. n is large and $np < 5$, so $X \sim Po(np)$, i.e. $X \sim Po(0.8)$ approximately.

$$P(X < 3) = e^{-0.8} + 0.8\,e^{-0.8} + \frac{0.8^2}{2!}e^{-0.8}$$

$$= 0.953 \text{ (3 s.f.)}$$

(b) If X is number of people with the health problem in a sample of size n, $X \sim B(n, 0.01)$.
If $n > 50$, $X \sim P(0.01n)$ approximately.

$P(X \geq 1) = 1 - P(X = 0) = 1 - e^{-0.01n}$

So $P(X \geq 1) > 0.95 \Rightarrow 1 - e^{-0.01n} > 0.95$
i.e. $e^{-0.01n} < 0.05$

By trial and improvement, $e^{-2.99} = 0.0502 > 0.05$, $e^{-3} = 0.049 < 0.05$, so $0.01n = 3 \Rightarrow n = 300$. (Log theory could be used here.) The minimum number in the sample should be 300.

3 (a) $k = \dfrac{1}{10}$

(b) By symmetry, $E(X) = \frac{1}{2}(2 + 12) = 7$

$$E(X^2) = \frac{1}{10}\int_2^{12} x^2\,dx = \frac{1}{10}\left[\frac{x^3}{3}\right]_2^{12} = 57\tfrac{1}{3}$$

$Var(X) = 57\tfrac{1}{3} - 7^2 = 8\tfrac{1}{3}$

Standard deviation $= \sqrt{8\tfrac{1}{3}} = 2.887$ (3 d.p.)

$P(7 - 2.887 < X < 7 + 2.887)$
$= P(4.113 < X < 9.887)$
$= 0.1 \times (9.887 - 4.113)$
$= 0.58$ (2 d.p.)

4 (a) $X \sim N(3.5, 0.8^2)$

$$P(X < 3.0) = P\left(Z < \frac{3.3 - 3.5}{0.8}\right)$$
$$= P(Z < -0.25)$$
$$= 1 - \Phi(0.25)$$
$$= 1 - 0.5987$$
$$= 0.4013$$

(b) $P(X > 3.25) = 0.95$

$$P\left(Z > \frac{3.25 - 3.5}{\sigma}\right) = 0.95$$

Now $P(Z > z) = 0.95 \Rightarrow z = -1.645$

so $\dfrac{3.25 - 3.5}{\sigma} = -1.645$

$3.25 - 3.5 = -1.645\sigma$

$-0.25 = -1.645\sigma$

$\sigma = \dfrac{-0.25}{-1.645} = 0.151\ldots = 0.15$ (2 d.p.)

5 For the binomial distribution,

$E(X) = np = 10 \times 0.4 = 4$

$Var(X) = npq = 4 \times 0.6 = 2.4$

By the central limit theorem, since the sample size is large,

$\bar{X} \sim N\left(4, \dfrac{2.4}{60}\right)$, i.e. $\bar{X} \sim N(4, 0.04)$

$$P(\bar{X} < 3.5) = P\left(Z < \frac{3.5 - 4}{\sqrt{0.04}}\right)$$
$$= P(Z < -2.5)$$
$$= 0.0062$$

Practice examination answers

Probability and Statistics 2 (continued)

6 Let X be the number of trees with the infestation. Assuming that trees are infested independently, with probability p, $X \sim B(n,p)$

$H_0: p = 0.35$
$H_1: p > 0.35$

(a) If $p = 0.35$, $X \sim B(10, 0.35)$
At the 10% level,
H_0 is rejected if $P(X \geq 6) < 0.1$

Using cumulative binomial tables,
$P(X \geq 6) = 1 - P(X \leq 5)$
$= 1 - 0.9051 = 0.0949$

Since $P(X \geq 6) < 0.1$, H_0 is rejected.

There is evidence, at the 10% level, that the percentage of trees that are infested is greater than 35%, indicating that the trees should be felled.

(b) If $p = 0.35$, $X \sim B(30, 0.35)$
n is large such that $np = 10.5 > 5$ and $nq = 19.5 > 5$ so use normal approximation.
$npq = 10.5 \times 0.65 = 6.825$
so $X \sim N(10.5, 6.825)$ approximately.

At the 10% level, H_0 is rejected if $z > 1.282$

Test $x = 14$.

Applying a continuity correction,
$z = \dfrac{13.5 - 10.5}{\sqrt{6.825}} = 1.148$

Since $z < 1.282$, H_0 is not rejected.
There is not enough evidence to say that the percentage of infected trees is greater than 35%. The trees should be treated with chemicals.

7 (a) $\displaystyle\int_{\text{all }x} f(x)\,dx = 1$

so $1 = k\displaystyle\int_0^1 (x - x^3)\,dx$

$= k\left[\dfrac{x^2}{2} - \dfrac{x^4}{4}\right]_0^1$

$= k\left(\dfrac{1}{2} - \dfrac{1}{4}\right)$

$= \dfrac{1}{4}k$

$\therefore \quad k = 4$

(b) $\mu = E(X) = \displaystyle\int_0^1 xf(x)\,dx = 4\int_0^1 x(x - x^3)\,dx$

$= 4\displaystyle\int_0^1 (x^2 - x^4)\,dx$

$= 4\left[\dfrac{x^3}{3} - \dfrac{x^5}{5}\right]_0^1$

$= 4\left(\dfrac{1}{3} - \dfrac{1}{5}\right)$

$= \dfrac{8}{15}$

(c) $P(X < 0.5) = 4\displaystyle\int_0^{0.5} (x - x^3)\,dx$

$= 4\left[\dfrac{x^2}{2} - \dfrac{x^4}{4}\right]_0^{0.5}$

$= 0.4375$

(d) $E(X^2) = \displaystyle\int_0^1 x^2 f(x)\,dx = 4\int_0^1 x^2(x - x^3)\,dx$

$= 4\displaystyle\int_0^1 (x^3 - x^5)\,dx$

$= 4\left[\dfrac{x^4}{4} - \dfrac{x^6}{6}\right]_0^1$

$= \dfrac{1}{3}$

$\text{Var}(X) = E(X^2) - \mu^2$

$= \dfrac{1}{3} - \left(\dfrac{8}{15}\right)^2$

$= 0.0489$ (3 s.f.)

Probability and Statistics 2 (continued)

8 (a) $P(X < 16) = P(X \leq 15) = 0.566$

(b) Let the number in stock be k.
We require $P(X > k) < 0.05 \Rightarrow P(X \leq k) \geq 0.95$

From tables, $P(X \leq 22) = 0.9418 < 0.95$
$P(X \leq 23) = 0.9633 > 0.95$

So there should be 23 copies in stock to meet the demand.

Practice examination answers

Mechanics 2

1 (a) Using $y = x\tan\theta - \dfrac{gx^2}{2v^2}(1+\tan^2\theta)$

gives $2.5 = 20\tan\theta - \dfrac{9.8 \times 20^2}{2 \times 25^2}(1+\tan^2\theta)$

$3.136\tan^2\theta - 20\tan\theta + 5.636 = 0$

$\tan\theta = 6.082\ldots$ or $\tan\theta = 0.2954\ldots$

$\tan^{-1}(0.2954\ldots) = 16.5°$

so a possible angle of projection is $16.5°$.

(b) Vertically: using $s = ut + \tfrac{1}{2}at^2$ gives

$-0.5 = 25\sin 16.5° \, t - 4.9t^2$

$4.9t^2 - 25\sin 16.5° \, t - 0.5 = 0$

$t = 1.516\ldots$ or $t = -0.067\ldots$

since $t > 0$

$d = 25\cos 16.5° \times 1.516 - 20$

The distance is 16.3 m to 3 s.f.

2 (a) 3 cm (by symmetry).

(b) $\bar{y} = \dfrac{24 \times 2 - 4 \times 3}{20} = 1.8$ cm

(c)

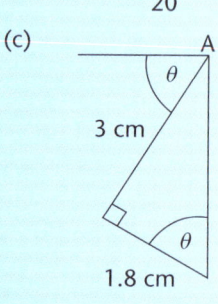

$\theta = \tan^{-1}\left(\tfrac{3}{1.8}\right) = 59.0°$

3 (a) Area rectangle : area triangle = 3 : 1

so mass of rectangle = 0.6 kg
mass of triangle = 0.2 kg

$\bar{x} = \dfrac{0.6 \times 7.5 + 0.2 \times 18\tfrac{1}{3}}{0.8} = 10.2$ cm

(b)

$15R_B = 10.2 \times 8 \Rightarrow R_B = 5.44$ N

$R_A = 8 - 5.44 = 2.56$ N

4 (a)

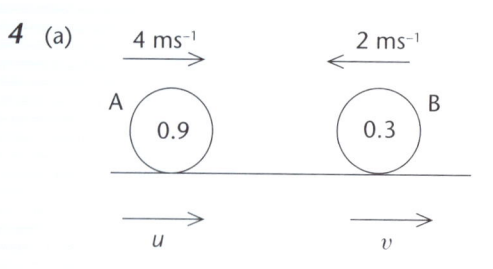

Conservation of momentum:

$0.9 \times 4 - 0.3 \times 2 = 0.9u + 0.3v$

so $3u + v = 10$ \hfill (1)

Newton's experimental law:

$v - u = \tfrac{1}{3} \times 6 = 2$ \hfill (2)

From (1) and (2) $4u = 8$

giving $u = 2$, $v = 4$.

The speeds of A and B after the collision are 2 ms⁻¹ and 4 ms⁻¹ respectively.

(b) Impulse = change in momentum

$= 0.3 \times 4 - 0.3 \times (-2)$

$= 1.8$ Ns

(c) Total KE before impact

$= \tfrac{1}{2} \times 0.9 \times 4^2 + \tfrac{1}{2} \times 0.3 \times 2^2 = 7.8$ J

Total KE after impact

$= \tfrac{1}{2} \times 0.9 \times 2^2 + \tfrac{1}{2} \times 0.3 \times 4^2 = 4.2$ J

Loss of mechanical energy = 3.6 J

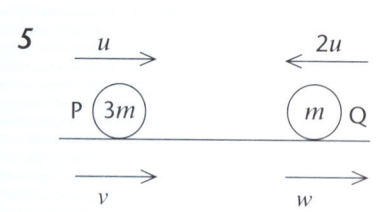

5

Conservation of momentum:

$3mu - 2mu = 3mv + mw$

giving: $3v + w = u$ \hfill (1)

Newton's experimental law:

$w - v = 3ue$ \hfill (2)

(1)−(2) gives $4v = u(1 - 3e)$

so $v = \dfrac{u}{4}(1 - 3e)$

$v < 0 \Rightarrow (1 - 3e) < 0$

$\Rightarrow e > \tfrac{1}{3}$

Practice examination answers

Mechanics 2 (continued)

6 (a) Before impact $5u$

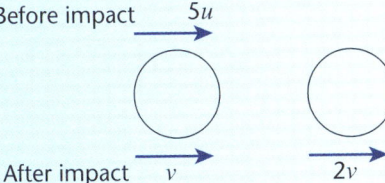

After impact v $\quad$ $2v$

Momentum before = momentum after

so $\quad 10mu = 2mv + 2mv$

$\Rightarrow 10u = 4v$

$\Rightarrow v = 2.5u$

(b) Impulse = change in momentum

$= 5mu - 0$

$= 5mu$.

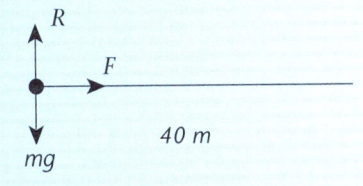

7

vertically: $\quad R = mg$

horizontally: $\quad F = \dfrac{m \times 15^2}{40}$

$\mu = \dfrac{F}{R} = \dfrac{15^2}{40 \times 9.8} = 0.57$

8

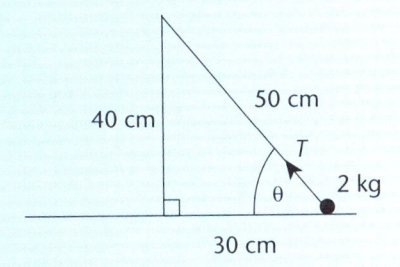

(a) Magnitude of acceleration $= \omega^2 r$

$= 3^2 \times 0.3$

$= 2.7 \text{ ms}^{-2}$

(b) Using $F = ma$: $\quad T \cos\theta = 2 \times 2.7$

so: $\quad T \times \tfrac{3}{5} = 5.4$

giving: $\quad T = 9 \text{ N}$

(c)

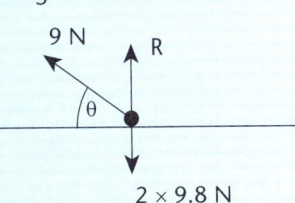

Vertically: $\quad R + 9 \sin\theta = 2 \times 9.8$

$R = 2 \times 9.8 - 9 \times 0.8 = 12.4$

$R = 12.4 \text{ N}$

9 (a) The tension is perpendicular to the direction of motion.

(b) The acceleration is not constant.

(c) Maximum speed occurs at lowest point.

Loss of PE = gain in KE

$mgl = \tfrac{1}{2}mv^2$

so, maximum speed is given by $v = \sqrt{2gl}$

10 (a)

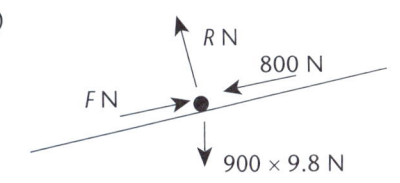

Resolving parallel to the plane:

$F = 900 \times 9.8 \times \tfrac{1}{15} + 800$

$= 1388$

The driving force is 1388 N.

(b) $P = Fv = 1388 \times 20 = 27760$

Power $\quad = 27.76 \text{ kW}$

(c) Using $F = ma$:

gives: $\quad 1388 - 800 = 900a \Rightarrow a = 0.65 \text{ ms}^{-2}$.

11 Momentum of ball before impact

$= 0.2 \times 20 = 4 \text{ Ns}$

Momentum of ball after impact

$= 0.2 \times 5 = 1 \text{ Ns}$

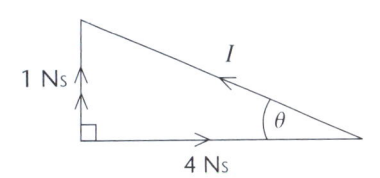

The magnitude of the impulse is given by:

$I^2 = 1^2 + 4^2 = 17 \Rightarrow I = \sqrt{17}$

$\tan\theta = \tfrac{1}{4} \Rightarrow \theta = 14.0°$

The impulse has magnitude $\sqrt{17}$ Ns at an angle of 14.0° above the horizontal.

Practice examination answers

Mechanics 2 (continued)

12 (a)

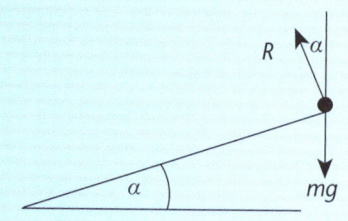

Vertically: $R \cos \alpha = mg$ (1)

Horizontally: $R \sin \alpha = \dfrac{m \times 20^2}{80}$ (2)

(2) ÷ (1) gives $\tan \alpha = \dfrac{20^2}{80 \times 9.8}$

$\Rightarrow \alpha = 27.0°$

(b)

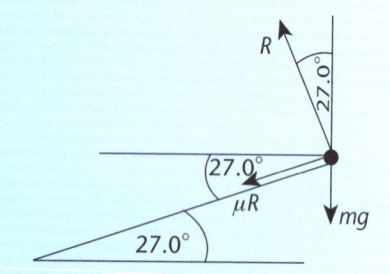

vert: $R \cos 27° = mg + \mu R \sin 27°$ (3)

horiz: $R \sin 27° + \mu R \cos 27° = \dfrac{m \times 28^2}{80}$ (4)

(4) ÷ (3) gives:

$\dfrac{\sin 27° + \mu \cos 27°}{\cos 27° - \mu \sin 27°} = \dfrac{28^2}{80 \times 9.8}$

rearranging gives:

$\mu = \dfrac{28^2 \cos 27° - 80 \times 9.8 \sin 27°}{80 \times 9.8 \cos 27° + 28^2 \sin 27°}$

$\Rightarrow \mu = 0.325$ to 3 s.f.

13 (a) Horizontally:

$T \sin 30° = \dfrac{0.2 \times 1.2^2}{0.4}$

$\Rightarrow T = 1.44$

The tension is 1.44 N.

Vertically:

$1.44 \cos 30° + R = 0.2 \times 9.8$

$\Rightarrow R = 0.713$

The reaction is 0.713 N to 3 s.f.

(b) Vertically: the new tension is given by:

$T \cos 30° = 0.2 \times 9.8 \Rightarrow T = 2.263$

Horizontally: $2.263 \sin 30° = 0.2 \omega^2 \times 0.4$

giving: $\omega = 3.76$

The minimum angular speed is 3.76 rad sec^{-1}.

Practice examination answers

Decision Mathematics 2

1 (a) Assume A uses Strategy W with probability p and strategy X with probability $(1 - p)$.

Expected gain for A is:

$4p + 6(1 - p)$ if B uses strategy Y.

$8p + 3(1 - p)$ if B uses strategy Z.

Optimal when

$4p + 6(1 - p) = 8p + 3(1 - p)$

$\Rightarrow -2p + 6 = 5p + 3$

$\Rightarrow p = \frac{3}{7}$

Assume B uses strategy Y with probability q and strategy Z with probability $(1 - q)$.

Expected loss for B is:

$4q + 8(1 - q)$ if A uses strategy W

$6q + 3(1 - q)$ if A uses strategy X.

Optimal when

$4q + 8(1 - q) = 6q + 3(1 - q)$

$\Rightarrow -4q + 8 = 3q + 3$

$\Rightarrow q = \frac{5}{7}$

The optimal strategy for A is to use strategy W with probability $\frac{3}{7}$ and strategy X with probability $\frac{4}{7}$.

The optimal strategy for B is to use strategy Y with probability $\frac{5}{7}$ and strategy Z with probability $\frac{2}{7}$.

(b) The value of the game is $5\frac{1}{7}$.

2 (a) The minimum row values are −4, −3 and −6. The maximum of these is −3 showing that the play-safe strategy for Richard is strategy B.

The minimum of the negatives of the column values are −4, −6 and −8. The maximum of these is −4 showing that the play-safe strategy for Judy is strategy X.

(b) $-3 + (-4) \neq 0$ so there is no saddle point.

(c) Comparing the row values shows that Richard's pay-offs are *always* less with strategy C than with strategy A. It follows that strategy C will not be a part of his mixed strategy.

(d) Assume Richard uses strategy A with probability p and strategy B with probability $(1 - p)$.

The expected gains for Richard based on the strategy used by Judy are:

X: $4p - 3(1 - p) = 7p - 3$

Y: $3p + 6(1 - p) = -3p + 6$

Z: $-4p + 8(1 - p) = -12p + 8$

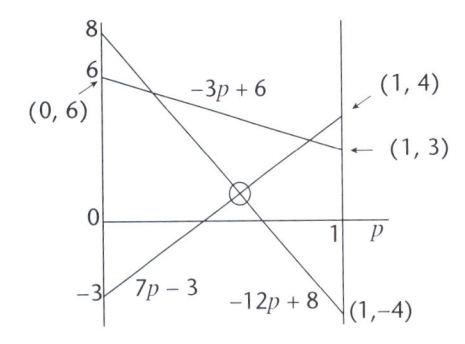

Optimal when $7p - 3 = -12p + 8$

$\Rightarrow p = \frac{11}{19}$

Richard's optimal strategy is to use strategy A with probability $\frac{11}{19}$ and strategy B with probability $\frac{8}{19}$.

Practice examination answers

Decision Mathematics 2 (continued)

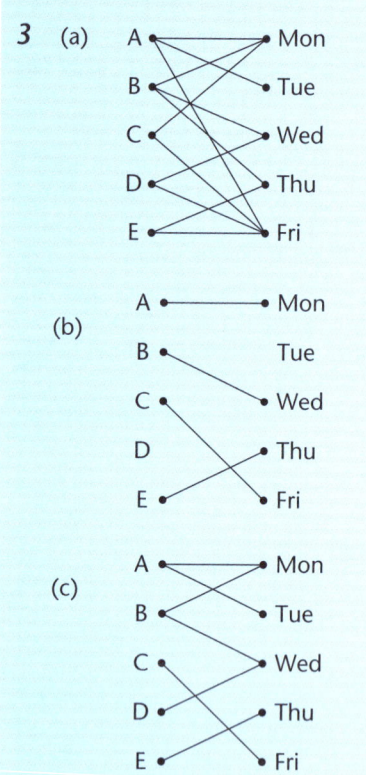

3 (a) Bipartite graph with A, B, C, D, E on left and Mon, Tue, Wed, Thu, Fri on right, with multiple connecting edges.

(b) Initial matching:
- A — Mon
- D — Thu
- E — Fri
(B, C unmatched to Tue, Wed)

(c) Matching after alternating path:
- A — Mon
- B — Tue (via crossing)
- D — Thu
- E — Fri

The alternating path is:
D – Wed – B – Mon – A – Tue.

The new matching is:
- A — Tue
- B — Mon
- C — Wed
- D — Thu (crossed with E)
- E — Fri

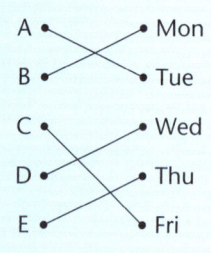

4

Stage	State	Action	Value	
1	G	GT	19	*
	H	HT	14	*
2	D	DG	Max(9, 19) = 19	
	E	EH	Max(17, 14) = 17	*
	F	FH	Max(21, 14) = 21	
3	A	AD	Max(12, 19) = 19	*
	B	BE	Max(25, 17) = 25	
		BF	Max(16, 21) = 21	*
	C	CD	Max(23, 19) = 23	
		CF	Max(22, 21) = 22	*
4	S	SA	Max(15, 19) = 19	*
		SB	Max(10, 21) = 21	
		SC	Max(20, 22) = 22	

The optimal minimax route is SADGT.
The maximum distance between supply points on the route is 19 km.

Index

acceleration 97
acceptance region 74
alternating path 121
alternative hypothesis 74
angle between lines 56
angular speed 97
area under curve 42

backward scan 118
binomial distribution 68
binomial expansion 15
binomial proportion test 76–77
bipartite graph 121

capacity 113
Cartesian form 25
census 71
central limit theorem 72
centre of mass 90–92
centripetal acceleration 97
centripetal force 98
chain rule 31
change of variable 43
circular motion 97–100
coefficient of restitution 94
collisions 93
composite function 19
compound angle identities 28
conical pendulum 98
connected rates of change 31
conservation of momentum 93
continuity correction 68
continuous random variables 64–65
coordinate geometry 25–26
coplanar forces 88
critical path 118
critical path analysis 117–118
critical region 74
critical value 74–75
cut 113

decimal search 49
definite integrals 41, 44
differential equations 46–47
differentiation 31–33
domain 18
dominance argument 112
double angle identities 29
dummy activity 117
dynamic programming 125–127

earliest event time 117
energy 99
equilibrium 87
errors (Type I and II) 75
estimation 71
events 117–118
expectation 65
exponential decay 47
exponential function 22–23, 32, 39
exponential growth 47

flow augmenting path 114
flows in networks 113
force 100
forward scan 117

game theory 109–112
gravitational potential energy 99

half-angle identities 29
Hungarian algorithm 122
hypothesis tests 74

identities (trigonometric) 28–29
image 18
implicit functions 37
improper fractions 15
impulse 95
integrals leading to logarithmic functions 41
integration 39–47
integration by parts 44
integration by substitution 43
inverse function 19, 28
iteration 50

joule 99

kinetic energy 99

lamina 91
latest event time 117
leaning ladders 89
logarithmic functions 23, 32, 41
logarithms 23

magnitude 53
matching improvement algorithm 121
matchings 121
maximin 126

maximum flow–minimum cut theorem 113
mean 65
mechanical energy 99
minimax 126
minimax point 110
mixed strategy 111
modulus function 21
momentum 93

networks 113–115
Newton's experimental law 94
normal approximations 68–69
normal distribution 64
null hypothesis 74
numerical integration 51

opportunity cost matrix 122

pairs of lines 56
parallel vectors 54
parametric differentiation 36
parametric form 25
partial fractions 15, 42
pay-off matrix 109
perpendicular vectors 55
play-safe strategies 110
Poisson approximation 68
Poisson distribution 67
Poisson mean test 77–78
population 71
population parameters 71
position vector 54
potential energy 99
power 100
precedence table 117
probability density function 65
product rule 35
projectiles 84
proper fractions 15
pure strategy 111
Pythagorean identities 28

quotient rule 35

radial acceleration 97
random number tables 71
random sample 71
range of a function 18
rational expressions 14, 42
reciprocal function 41
rejection region 74

Index

resource histogram 119
resource levelling 119
resultant vector 53

saddle point 110
sampling distributions 72
sampling techniques 71
saturated edge 114
scalar product 55
scheduling 118
separation of variables 46
series 15
sign change 49
significance level 74
significance test 74
simple random sample 71
Simpson's rule 51
sink vertex 113
skew lines 56
source vertex 113
stable solution 110
stages 125
standard deviation 65, 72
standard error of the mean 72
start vertex 117
stationary points 25, 36
statistical approximations 67–69
stratified sampling 71
super sink 114
super source 114
systematic sampling 71

terminal vertex 117
test statistic 74
test value 74
total float 118
trajectory formula 86
transformations 21
trigonometric equations 30
trigonometric functions 27, 34, 39–40
trigonometric identities 28

trigonometry 27–30
Type I error 75
Type II error 75

unbiased estimate 71
unit vector 54

variance 65
vector equation of a line 55
vectors 53–57
vertex 113, 117
volume of revolution 45

work 99

zero-sum game 109
z-tests 75